Emergence, Complexity and Computation

Volume 20

About this Series

The Emergence, Complexity and Computation (ECC) series publishes new developments, advancements and selected topics in the fields of complexity, computation and emergence. The series focuses on all aspects of reality-based computation approaches from an interdisciplinary point of view especially from applied sciences, biology, physics, or Chemistry. It presents new ideas and interdisciplinary insight on the mutual intersection of subareas of computation, complexity and emergence and its impact and limits to any computing based on physical limits (thermodynamic and quantum limits, Bremermann's limit, Seth Lloyd limits...) as well as algorithmic limits (Gödel's proof and its impact on calculation, algorithmic complexity, the Chaitin's Omega number and Kolmogorov complexity, non-traditional calculations like Turing machine process and its consequences,...) and limitations arising in artificial intelligence field. The topics are (but not limited to) membrane computing, DNA computing, immune computing, quantum computing, swarm computing, analogic computing, chaos computing and computing on the edge of chaos, computational aspects of dynamics of complex systems (systems with self-organization, multiagent systems, cellular automata, artificial life,...), emergence of complex systems and its computational aspects, and agent based computation. The main aim of this series it to discuss the above mentioned topics from an interdisciplinary point of view and present new ideas coming from mutual intersection of classical as well as modern methods of computation.Within the scope of the series are monographs, lecture notes, selected contributions from specialized conferences and workshops, special contribution from international experts.

More information about this series at http://www.springer.com/series/10624

Andrew Adamatzky · Genaro J. Martínez
Editors

Designing Beauty:
The Art of Cellular Automata

 Springer

Editors
Andrew Adamatzky
Unconventional Computing Centre
University of the West of England
Bristol
UK

Genaro J. Martínez
Unconventional Computing Centre
University of the West of England
Bristol, Avon
UK

ISSN 2194-7287   ISSN 2194-7295   (electronic)
Emergence, Complexity and Computation
ISBN 978-3-319-27269-6   ISBN 978-3-319-27270-2   (eBook)
DOI 10.1007/978-3-319-27270-2

Library of Congress Control Number: 2015958339

This Springer imprint is published by SpringerNature
The registered company is Springer International Publishing AG Switzerland

Preface

Cellular automata are regular uniform networks of locally-connected finite-state machines.They are discrete systems with non-trivial behaviour. Cellular automata are ubiquitous. They are mathematical abstractions of computation, models of physical, chemical and livings systems, and architectures of massive-parallel processors. Cellular automata generate patterns. These patterns feed our visual thinking. They help us to discover novel properties of spatially extended systems. They aid us in design of parallel algorithms. Also the patterns excite us. They fuel our imagination and guide us on the trips into depth of unknown, in the galaxies of the Computing Universe.

Science aims for results. Art is driven by process. A kaleidoscope of colourful complex patterns produced by apparently simple rules is where science merges with art, and art becomes part of science. World leading mathematicians, computists, physicists, and engineers brought together marvellous, entertaining and often esoteric configurations generated by cellular automata with a rich family of local state-transition rules, including totalistic, cyclic and reversible automata, majority vote, asynchronous, excitable and lattice-gas automata. The automata evolve on orthogonal and hexagonal lattices, Penrose tilings, geodesic grids and hyperbolic planes. Many works are produced using Conway's Game of Life automata and their modifications: Larger than Life, Life without Death, enlightened Game of Life. Computational potential of cellular automata is illustrated by snapshots of evolving counters, automata solving firing squad synchronisation and Prisoner's dilemma problems, self-reproduction, and a design of universal Turing machine implemented in the Game of Life.

Configurations produced by cellular automata help us to get an insight into the mechanics of pattern formation, propagation and interaction in natural systems: heart pacemaker, bacterial membrane proteins, chemical reactors, water permeation in soil, compressed gas, cell division, population dynamics and non-trivial collective behaviour, reaction-diffusion media and self-organisation. Examples of real architectural forms, ornamental systems and floor tilings presented in the book bridge virtual beauty of local transitions rules with aesthetic appealing of physical objects

VI

generated by the rules. Many of the cellular automata art works have been shown at major art exhibitions, installations and performances; others are newly born and awaiting for their fame to come.

The book offers in-depth insights and first-hand working experiences into production of art works, using simple computational models with rich morphological behaviour, at the edge of mathematics, computer science, physics and biology. We believe the works presented will inspire artists to take on cellular automata as their creative tool and will persuade scientists to convert products of their research into the artistic presentations attractive to general public.

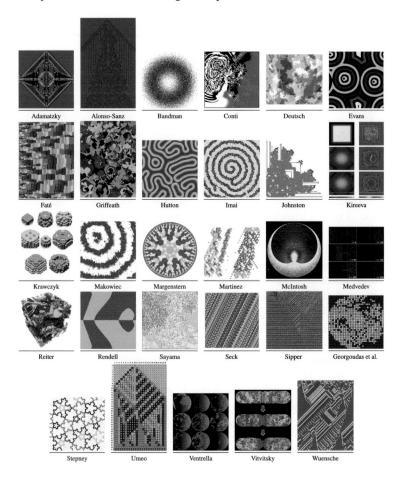

Andrew Adamatzky, Bristol
Genaro Martinez, Mexico City
January, 2016

Contents

Self-Organizing Two-Dimensional Cellular Automata: 10 Still Frames

David Griffeath

A favorite topic of my research in the 1980s and 1990s was pattern formation of interacting particle systems and cellular automata (CA) started from random initial states, especially in two dimensions. Together with my students, and later colleagues, Bob Fisch, Janko Gravner, and Kellie Evans, I studied a wide variety of local lattice evolutions started from noise. With the advent of personal computers it became possible to interact with simulations of these algorithms in order to observe system behavior and sometimes gain insight leading to rigorous mathematical results. Fisch and I began developing WinCA [52], a user-friendly environment for experimentation with 2D rules, in the early 1990s. By the fall of 1994, with the World Wide Web well established, I launched my Primordial Soup Kitchen website [69][1] to introduce our research to a more general audience through graphics, animations, commentary, and other resources. In particular, over a two-year span I offered weekly "soup recipes": colorful snapshots of our experiments with companion descriptions.

Twenty years have passed since the Primordial Soup Kitchen was really cooking. WinCA is now incompatible with the latest versions of Windows, and I have retired. However, when the editors kindly invited me to contribute to this Atlas, it occurred to me that the fairly recent XP machine in my basement still runs the simulator, with video memory and processing speed one could only dream about in the mid 1990s. So I am delighted to offer still frames of 10 of my favorite soups from the website on much larger arrays, many with new color palettes that I hope the viewer will find passably artistic. Each image is accompanied by a brief explanation of how it was generated and one or more references for mathematical background.

Before turning to commentary about the individual graphics, let me mention some common features. The size of each of the 10 systems, and resolution of the corresponding image, is 1600×1200 cells. Every evolution starts from a random initial

D. Griffeath
University of Wisconsin, Madison, USA
e-mail: griffeat@math.wisc.edu

[1] http://psoup.math.wisc.edu

© Springer International Publishing Switzerland 2016
A. Adamatzky and G.J. Martínez (eds.), *Art of Cellular Automata*,
Emergence, Complexity and Computation 20,
DOI: 10.1007/978-3-319-27270-2_1

configuration over the array, with prescribed densities for the possible states. The first 5 depict excitable dynamics, with a latent state labeled 0, and additional excited or refractory states $1, \ldots, N-1$, as specified in each case. These 5 graphics display the configuration of the resulting N-color CA at a specified time t when pattern formation is either in process or complete, using a suitable palette intended to highlight structure. Graphic 10 also shows a system with N states at time t, where N and t are both large, using a palette of N colors to represent the cell states. Graphics 6-9 are generated by CA rules with only two states, 0 (= empty) and 1 (=occupied). For these, I display successive updates by means of a large periodic palette in order to enhance the visualization of dynamic self-organization. Namely, I give empty cells a fixed background color while coloring occupied cells with the cyclically coded time of last transition from 0 to 1. For systems which evolve to a final fixed configuration or limit cycle over a long period of time, this level-set scheme produces a rich 'fossil record' that can be both beautiful to behold and mathematically illuminating. It should also be noted that graphics 5 and 10 are generated by algorithms with random ingredients, i.e., probabilistic cellular automata (PCA). All the others depict deterministic CA rules. Finally, in each case I specify the boundary conditions used to update cells at the array edges; typically this will be wrap (periodic), but in a few cases 0 (neighbors outside the array have fixed value 0).

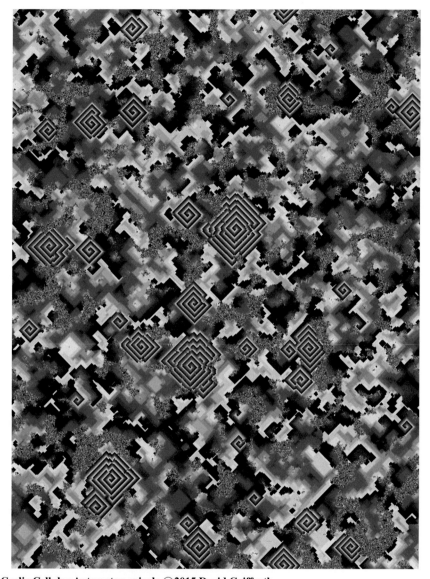

Cyclic Cellular Automaton spirals ⓒ2015 David Griffeath
Cyclic Cellular Automata (CCA) exhibit complex self-organization by iteration of an extremely simple update rule. Using a cyclic color wheel, the state of each cell advances to the next in the cycle if and only if it sees a sufficient representation of that next color within its prescribed local neighborhood. 'Enough' is measured by a threshold parameter. The basic CCA, with a range 1 diamond neighborhood consisting of a cell and the four lattice sites at distance 1 (N, S, E, W), and with threshold 1, was introduced in [54]. For popular science accounts, see [37] and [53]. An in-depth study of CCA with larger neighborhoods and arbitrary thresholds appears in [55]. Here we show the basic CCA with 16 colors, started from a uniform random distribution over the colors, and with wrap boundary conditions. The palette is comprised of a range of colors from black (state 0), through orange, to light yellow (state 15). Our snapshot is taken after 360 updates, when pattern formation exhibits spiral formation within disordered cyclic waves as well as patches of residual "noise" from the initial configuration.

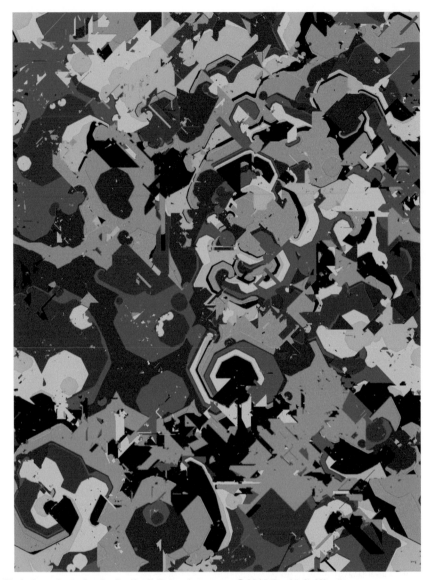

Turbulent clustering in Cyclic Cellular Automata ©2015 David Griffeath

What happens in CCA dynamics when the threshold is too high for spiral formation? Roughly speaking, the system tries to make spiral centers, but these would-be centers are unstable. As a result, when a vortex unravels it leaves behind a smattering of debris that can act as a seed for the formation of new wave fronts. This debris is a permanent and crucial part of the dynamics; the resulting "turbulence" is characterized by ever-increasing length scales. Details appear in [55]. For this graphic, the neighborhood of a cell is range 2 box, by which we mean a 5×5 lattice square centered at that cell, the threshold is 5, and there are 8 colors. Again, we start from a uniform random mix of the colors, and use wrap boundary conditions. After 2,000 updates there is an extremely complex mix of large-scale wave activity, incomplete spirals, linear and polygonal forms, and tiny, disordered debris. Our palette for this image was inspired by the late work of Wassily Kandinsky.

Greenberg-Hastings spiral pairs ©2015 David Griffeath

Perhaps the simplest cellular automaton rule for an excitable medium is known as the Greenberg-Hastings Model (GHM). As for CCA, a prescribed number of colors N are arranged cyclically in a color wheel, and each can only advance to the next, the last cycling to 0. Every update, cells change from color 0 (latent) to 1 (excited) if they have at least threshold 1s in their neighbor set. All other colors $2, \ldots, N-1$ (refractory) advance automatically. The Belousov-Zhabotinsky oscillating chemical reaction is a Petri dish experiment that produces spiral waves strikingly similar to those generated by the Greenberg-Hastings Model. In contrast to CCA, both these virtual and physical waves come in spiral pairs, sometimes called ram's horns. Again, see [55] for an extensive account of GHM dynamics. Our still frame of prototypical GHM self-organization has a range 4 box neighborhood, threshold 8, and 8 colors. Once more we start by coloring sites randomly and use wrap boundary conditions. By time 100 the entire array has settled into a locally periodic pattern driven by stable spiral pairs. For fun, I have chosen a high-intensity 8-color palette derived from one of my favorite prints by Keith Haring.

Greenberg-Hastings metastability ©2015 David Griffeath
How many colors can support Greenberg-Hastings model self-organization if the initial mix is not uniform, but instead optimally chosen to induce spiral formation? It turns out that in the basic case (range 1 diamond, threshold 1), a mix of about 61% 0s and 39% 2s, together with a very small smattering of excited states, can nucleate with 180 colors on grids containing nearly two million cells. This graphic shows surviving fronts of excitation in the process of forming stable spiral cores after 200 updates. The precise probability distribution over colors for the initial random mix is (.6095, .0005, .39, 0, 0, ..., 0) and the boundaries wrap. The palette has black as the latent state 0, fire red as the excited state 1, and then a spectrum for the refractory states through yellow, green, cyan, and purple, to dark blue (179). For more about the mathematics behind these dynamics, which is intimately connected with percolation theory, see [56].

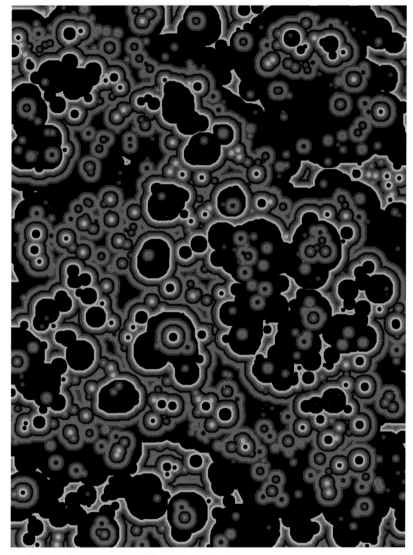

Digital Boiling ©2015 David Griffeath

This graphic depicts one variety of ring dynamics that can be encoded as an excitable PCA. There are 8 states: 0 (latent) is black, 1 (excited) is bright blue, and 6 additional bright colors with good contrast represent the refractory stages. At each time step, latent sites become excited spontaneously with a small probability, $p = .0001$ in this instance. In addition, a latent site always become excited if there is at least one excited cell within its neighborhood, a range 6 lattice ball here (neighbors of a site are all within Euclidean distance 6.5). With each update, excited cells become refractory and refractory cells cycle until they again become latent. The resulting evolution consists of expanding rings that nucleate at random locations and annihilate upon collision, a kind of Digital Boiling. Started by exciting each site with probability .01, and leaving the remainder of the array latent, after 600 updates the system settles into the apparent 'bubbling' equilibrium shown in our still frame. For no particular reason, 0 boundary conditions are used in this case; wrap produces indistinguishable behavior away from the edges of the array. A thorough and elegant mathematical study of ring dynamics appears in [65].

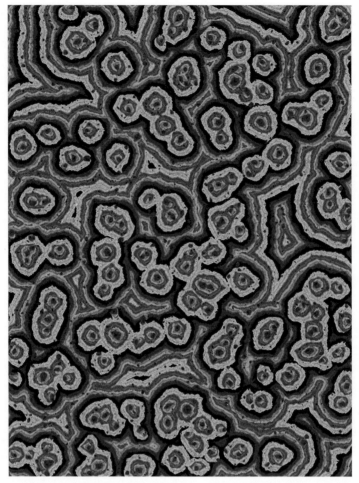

Bootstrap Percolation ©2015 David Griffeath

Start with a random density p of occupied sites (1s on a background of 0s). At each update an empty site becomes occupied if it sees at least two previously occupied sites among its four nearest neighbors (N, S, E, W). Once occupied, any site remains so thereafter. Under these dynamics an isolated rectangle of occupied sites cannot grow, but any occupied region that is not convex must continue to grow. This is basic Bootstrap Percolation, a much studied model for nucleation and metastability in statistical physics; see [11] for an introduction to the mathematics. Here we feature another bootstrap rule with slightly larger neighborhood and threshold: an empty cell must see 4 occupied sites among its 8 nearest neighbors to become occupied. Like the basic case, our variant is critical, meaning that finite configurations of 1s are convex-confined and can only spread indefinitely if they repeatedly get a boost from external occupied cells. Consequently, with the aid of a level set representation we are able to convey visually the complex interaction between nucleating droplets. Our initial seeding has $p = .91$, a large enough density for the nucleating "droplets" of 1s to cover the lattice by time 1,000. We use the 0 boundary condition, though wrap would give an almost identical result. Our cyclic palette for the level-set representation consists of 201 colors (the most supported by WinCA in this mode). 0 (= empty) is dark blue, nowhere to be seen once the array is filled with 1s, and the remaining 200 hues strike a balance between gradient and high-contrast transitions.

Majority Vote ©2015 David Griffeath
This graphic features the range 10 box version of Majority Vote, with wrap boundary conditions, in which a cell switches allegiance (Democrat vs. Republican, 'for' vs.'against', etc.) whenever the majority of its 441 neighbors hold the opposite opinion. Starting from a 50-50 completely random mix of the two opinions, within a few updates the noise self-organizes into well-separated components of 0s and 1s. Over time, the minimal distance between any two distinct components of the same type increases, and any sufficiently small component completely surrounded by the opposite type vanishes. Because of wrap-around, the result is a kind of random maze on a torus. If there is a loop through 1s connecting the top to the bottom, and another loop of 1s connecting the left side to the right, then together these guarantee that 1s constitute a background for the entire array, in which case the level set coloring scheme yields spectacular graphics if continued until a final state of all 1s is reached. It only takes a few restarts to obtain such a scenario, as in our still frame at time 1,908. Using the level set representation, our graphic conveys the process of self-organization from the "big bang" until 1's cover the lattice. Even though the visual effect is rather different, the palette used is exactly the same as for the previous example. Additional discussion of Majority Vote appears in [68].

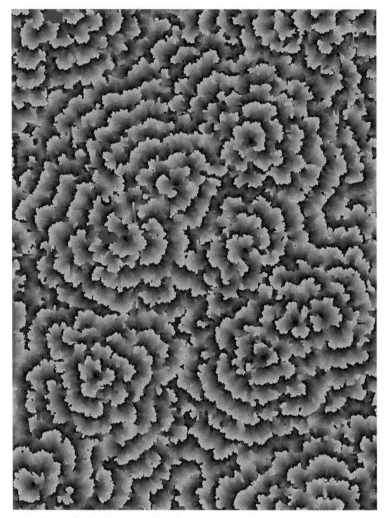

Life without Death ©2015 David Griffeath

An empty cell becomes occupied if there are exactly 3 occupied cells in its range 1 box neighborhood, and once a cell is occupied it remains so forever. This rule is like Conway's celebrated Game of Life, but without death. Of course one expects the occupied region to grow over time, but how does it evolve from finite initial configurations of occupied cells? E.g., starting from a lattice ball of radius 50, vertical and horizontal ladders emerge, evolving by a kind of weaving pattern that seems to outrun the more complex growth elsewhere, and therefore might be expected to continue indefinitely. However, later on parasites arise on the edges of the ladders that travel even faster until they reach the ladder tips and stop the weaving in its tracks. It turns out that Life without Death (LwoD) supports such a complex collection of local interactions that it can emulate all the basic logic gates and thereby perform algorithmic computation [70]. Here we offer an LwoD graphic in the same vein as the previous two for Bootstrap Percolation and Majority Vote. Start with a small random smattering of occupied cells, with density .005, and use wrap boundary conditions. By time 3,000 the evolution has achieved its final configuration. A level-set representation with 0s in dark brown, and 1s in 200 shades ranging from near black through light brown to white, evokes a botanical feel, with forms somewhat akin to Ginkgo leaves. The occasional horizontal and vertical linear structures are ladder segments.

Mandala Batik ©2015 David Griffeath
During the early 1990s, in collaboration with Kellie Evans, I introduced and studied a parameterized family of two-color CA rules known as Larger than Life (LtL). In this straightforward generalization of Conway's Game, a 1 is born at site x if the number of 1's in the neighborhood of x lies in the interval $[b_1, b_2]$, while a 1 survives at site x if the number of 1's in the neighborhood lies in $[d_1, d_2]$. For simplicity, we take the neighbor set to be a range r box. Thus Conway's Game is the case $r = 1$, $b_1 = b_2 = d_1 = 3$, and $d_2 = 4$. LtL has a remarkably rich phase space, with many qualitatively different kinds of long-term, complex pattern formation. For an introductory account with a couple of pictures, see the final section of [68]; many more wonderful Larger than Life stories appear in [44]. Our graphic here depicts a regime not mentioned in either paper. The parameters are $r = 13$, $b_1 = 10$, $b_2 = 100$, $d_1 = 100$, and $d_2 = 320$. Start from noise, i.e., a symmetric random mix of 0s and 1s. By time 349, again using the level-set representation, the entire system has settled into a stable periodic orbit, with fixed values at most sites, but also some isolated small blinkers (mostly of period 2 or 4). Our palette has 0 as black and a spectrum of 200 cheery colors to capture the pattern formation of 1s. When I showed a picture of this rule to an art class at Dartmouth College many years ago, the nucleating target patterns reminded some students of mandala symbols they had studied. For me, the overall effect evokes the wonderful designs of batik and Australian indigenous art. An amusing side note: by choosing 0 boundary conditions, this rule creates its own frame, with a noisy edge and then slightly irregular alternating black and red stripes before settling into the dominant pattern.

Stepping Stone probabilistic CA ©2015 David Griffeath
This PCA derives from work by population geneticist Sewall Wright in the early 1940s, as a prototype for selectively neutral competition. We start from a completely random configuration, using
a rainbow spectrum of 256 colors (the most supported by WinCA in general). Then we impose an
extremely simple iterative rule, with wrap boundary conditions: flip a fair coin at each site of the
array and if it come up 'heads,' give that site the color of a randomly chosen one of its four nearest
neighbors; otherwise the color at the site is unchanged. Over time, increasingly large but highly
irregular regions of solid color are formed until, eventually, one "species" takes over the entire
array. (Coin flips at each site introduce asynchrony to eliminate two-color checkerboard patterns
that would otherwise emerge.) Our still frame shows the dynamics after 20,000 updates. A precise
mathematical description of this exotic diffusive clustering is presented in [33] for a continuous-
time, two-color variant known as the voter model; see also [34] for historical background and mul-
titype results. The Stepping Stone model has a special place in my life's work since one version
was the first lattice interaction I ever simulated, on a Commodore 64 in the mid 1980s.

Is it Art or Science?

Andrew Wuensche

Is the art of cellular automata (CA) a legitimate subject of preoccupation? — yes, of course! Although the study of CA is an exercise in experimental dynamics by computer algorithms on top of which mathematical theories and conjectures are superimposed, the graphic representations themselves confer intuitive subjective impressions that are inescapably art. To the simple art lover these are intriguing immediate images which imagination may strive to interpret or merely accept. To the CA theorist and practitioner the "art" is imbued with layers of deeper meaning, just as Zen art can be experienced either on the surface or by the Zen master.

In 1989 I discovered algorithms to run a CA "backwards" to compute the predecessors of any state, and realised these states could be joined up into a graph with the topology of "trees" rooted on "attractor" cycles. The challenge was to represent the graphs — initially I simply hand drew them by pencil on paper, designing workable graphic conventions. My previous career as an architect made the drawing easy though tedious. Translating hand drawing into automatic, immediate, computer drawing followed, and although the charm of hand drawing was lost, computer drawing allowed for much larger systems and provided a striking and powerful aesthetic, including a dynamic, organic, feel were "trees" grow outwards from "attractors" as the "backwards" computation process unfolds.

The theory and results of these experiments were published as a book "The Global Dynamics of Cellular Automata, An Atlas of Basin of Attraction Fields of One-Dimensional CA" ("the Atlas" for short) in the Santa Fe Institute's "Studies in the Sciences of Complexity" 1992 [171]. The book included a diskette with the software for generating space-time patterns and basins of attraction, and color plates of basins of attraction images. The book's letter size format prompted some to call it a coffee table book, yet I was keen on its scientific significance, not its art, although I admit to putting time and effort into making the images as aesthetically pleasing as possible — to myself.

A. Wuensche
Discrete Dynamics Lab, UK
e-mail: andy@ddlab.org

© Springer International Publishing Switzerland 2016
A. Adamatzky and G.J. Martínez (eds.), *Art of Cellular Automata*,
Emergence, Complexity and Computation 20,
DOI: 10.1007/978-3-319-27270-2_2

The consequence of the Atlas became in time the program Discrete Dynamics Lab — "Tools for researching CA, Random Boolean Networks, multi-value Discrete Dynamical Networks, and beyond" [166, 170]. DDLab is the generator of many types of images, and I am very aware and unapologetic that these images must have aesthetic value as well as being diagrams representing scientific concepts and data.

The images presented here were created by DDLab's open source software according to the various parameters and methods described in the captions. As well as basins of attraction, these images include space-time patterns, or one-time snapshots, which is the type of image most commonly associated with CA, whereas basins of attraction are a rarer breed, generated by DDLab and few other software resources. A space-time pattern is actually a single trajectory that one can imagine starting at a leaf, then moving down branching sub-trees to the trunk – the attractor of the basin of attraction. In a better metaphor, the trajectory starts as a single raindrop falling on a landscape with hills and lakes, and flows downhill along streams and rivers to one particular lake — the attractor. Had the raindrop fallen at a slightly different place, it may well have ended up in a different lake or attractor. DDLab is able to organise the whole state-space into the set of basins of attraction. In a CA, where time advances in discrete steps, and space is course-grained to make a discrete lattice, the trajectory turns into a sequence of discrete images, like the frames of a movie. As time proceeds, initial patterns may self-organise into "emergent" forms. This is what underlies the aesthetic appeal of space-time patterns, or their one-time snapshots, though seeing the actual movie is even better!

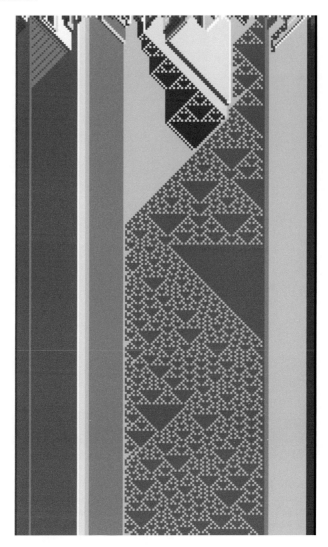

1D space-time pattern of a probabilistic majority rule. ©2014 Andrew Wuensche.
A 1D space-time pattern from a random initial state. $n=150$, $v=8$, $k=4$, showing the evolution of a probabilistic majority rule, where each lookup table output is preset according to the frequency of values in its neighborhood. This type of rule was suggested by Lee Altenberg for binary CA, but is extended here to multi-value, in this case 8 values or colors. The method is described in "Exploring Discrete Dynamics" [170]. The resulting dynamics stabilise into zones of order, but we also see Sierpinski triangle patterns from interactions between specific color pairs. The image was taken from the DDLab website [166], http://uncomp.uwe.ac.uk/wuensche/multi_value/ddlab_multi_value.html.

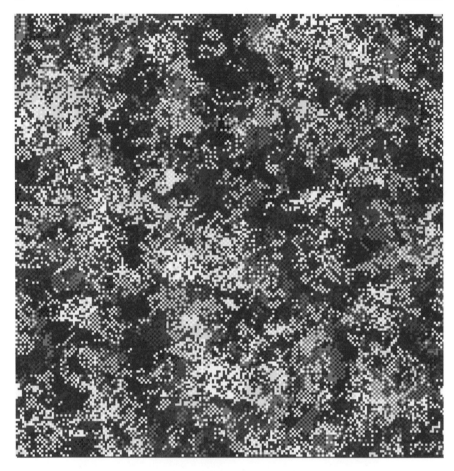

2D space-time pattern of a probabilistic majority rule. ©2014 Andrew Wuensche.
A 2D space-time pattern from a random initial state. $n=200 \times 200$, $v=8$, $k=4$, showing an evolved
snapshot applying a probabilistic majority rule, where each lookup table output is preset according
to the frequency of values in its neighborhood, as in the previous figure. The method is described
in "Exploring Discrete Dynamics" [170]. The resulting dynamics settles into swirling patches of
order and interaction. The image was taken from the DDLab website [166], `http://uncomp.`
`uwe.ac.uk/wuensche/multi_value/ddlab_multi_value.html`.

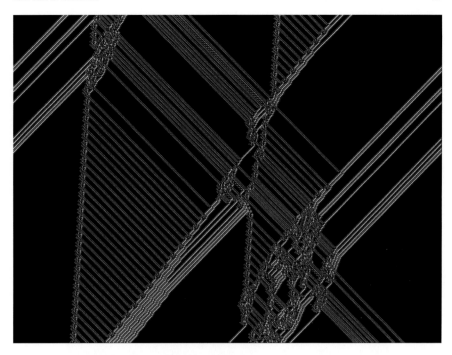

Long History of a Small Universe. ©2014 Andrew Wuensche.
About 600 time-steps of a 3000 time-step image called "Long History of a Small Universe", part of a set of images "Complexity in Small Universes" [168], presented by Chris Langton and Andrew Wuensche at an exhibition of art connected with science, entitled, "Objective Wonder: Data as Art", at the University of Arizona in March 1999. This is the space-time pattern of a binary 1D CA, n=900, k=5, rule (hex) 360a96f9, which self-organises a random initial state into mobile interacting configurations, gliders and glider-guns. Cells are colored according to neighborhood lookup instead of the value. This rule and similar complex rules were found by an automatic classification method [167]. The image was taken from the DDLab website[166], http://uncomp.uwe.ac.uk/wuensche/Exh2/Exh3.html.

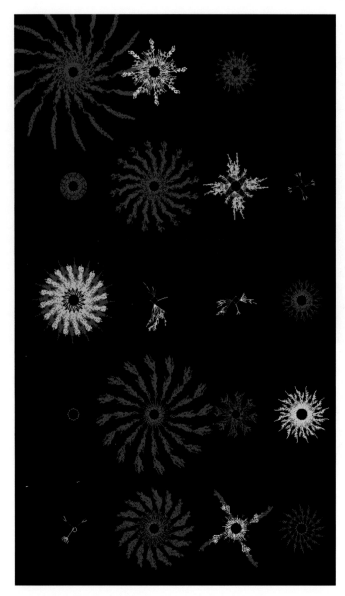

The basin of attraction field of a 1D binary complex CA. ©2014 Andrew Wuensche.
A basin of attraction (state transition graph) links unique states in state-space according to their transitions in time (time-steps). Because a state has one successor but multiple or zero predecessors, the dynamics is convergent like the flow in a river system, in other words with a tree topology. However, for finite systems a transition must cycle — the trees are rooted on periodic attractors. This represents how dynamics organise state-space, and provides a global perspective on CA [171]. This image represents the set of all the non-equivalent basins of attraction for the binary $k=5$ complex rule (hex) 6c1e53a8, $n=16$, and was part of the same set of images "Complexity in Small Universes" [168] in the art exhibition described in the previous figure. The image was taken from the DDLab website [166].

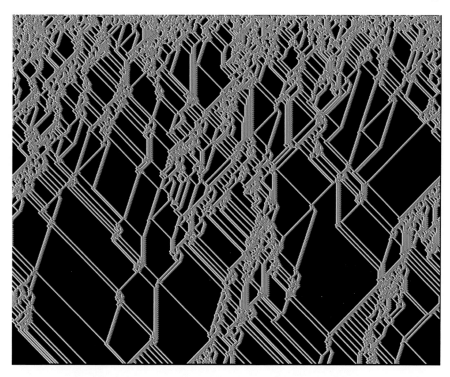

Space-time pattern of a 1D binary complex CA. ©2014 Andrew Wuensche.
This is the same 1D k=5 rule (hex) 6c1e53a8, as in the previous figure, but shows the space-time patterns for a much larger system size, n=900. This complex rule self-organises a random initial state into mobile interacting configurations, gliders and glider-guns. Cells here are colored according to neighborhood lookup instead of value. This rule and similar complex rules were found by an automatic classification method [167]. This was part of the same set of images "Complexity in Small Universes" [168] in an art exhibition described previously. The image was taken from the DDLab website [166], http://uncomp.uwe.ac.uk/wuensche/Exh2/Exh3.html.

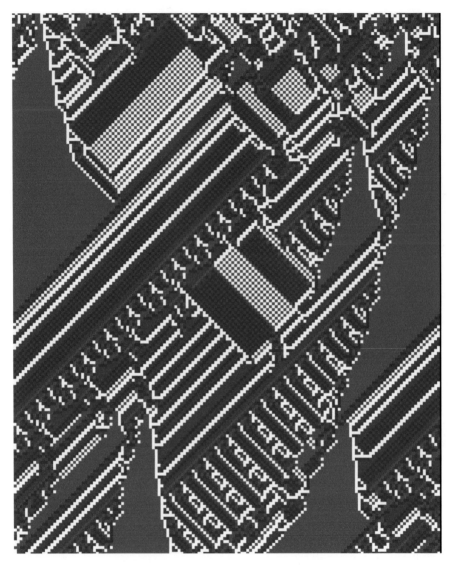

Space-time pattern of a 4-value 1D CA. ©2014 Andrew Wuensche.
A 1D space-time pattern from a random unbiased initial state. n=150, v=4, k=3, of the complex rule
(hex) 60ef5f1ceba19251e32260940d34f864. This rule and similar complex rules were found by an
automatic classification method [167, 169]. The evolution produces a diversity of quasi-stable and
complex patterns. The image was taken from the DDLab website [166], http://uncomp.uwe.
ac.uk/wuensche/multi_value/ddlab_multi_value.html

Basin of attraction of a binary totalistic rule. ©2014 Andrew Wuensche.
This is a reconstruction of one of 28 color plates that appeared in the book "The Global Dynamics of CA" by Wuensche and Lesser [171] in 1992. The reconstruction is taken from the DDLab website [166], http://uncomp.uwe.ac.uk/wuensche/gdca_images.html. The binary $k=5$, $n=19$, CA has the totalistic rule 10 (001010) in Wolfram's notation. The idea of basins of attraction were described in the introduction. This figure shows just one basin of attraction with 3369 nodes representing unique states, linked by directed arcs, forming trees rooted on the attractor cycle. Each tree is assigned a color, alternating between 4 colors. The direction of time is inward from the leaves (white dots) which have no predecessors, called "garden-of-Eden" states, towards the central period 10 attractor, then clockwise. In the DDLab software [166], we see the basin of attraction generated from the attractor outwards in real time, by computing the predecessors of each state with a backwards-in-time algorithm [171].

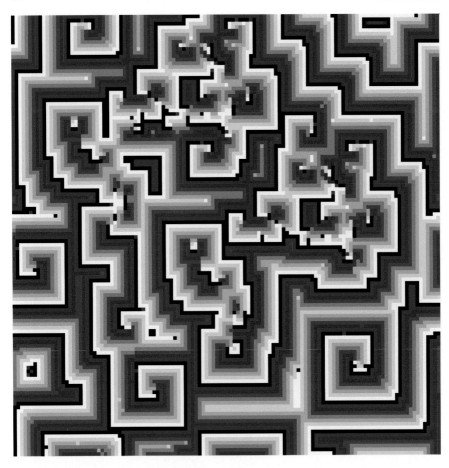

Reaction-Diffusion CA. ©2014 Andrew Wuensche.
Reaction-diffusion or excitable media dynamics can be generated with a type of CA which mimics Belousov-Zhabotinsky oscillating chemical reactions. The CA has 3 cell qualities: resting, excited, and refractory (or substrate, activator, and inhibitor). There is usually one resting type, one excited type, and one or more refractory types. A resting cell becomes excited if the number of excited cells in its neighborhood falls within the threshold interval. At each time-step, an excited cell type changes to refractory. A refractory cell type changes to the next refractory type (if there are more than one) and the final refractory type changes back to resting, completing the cycle. In this snapshot of an evolved state on a 122×122 square grid, v=8, k=8, the values 0=rest, 1=excited, 2 to 6 are refractory. The threshold interval is 1 to 6. The non-resting initial density is about 30% which is a critical parameter for the characteristic spirals to form. The image is taken from the DDLab website [166], http://uncomp.uwe.ac.uk/wuensche/update_dec04.html and the method is described in "Exploring Discrete Dynamics"[170] .

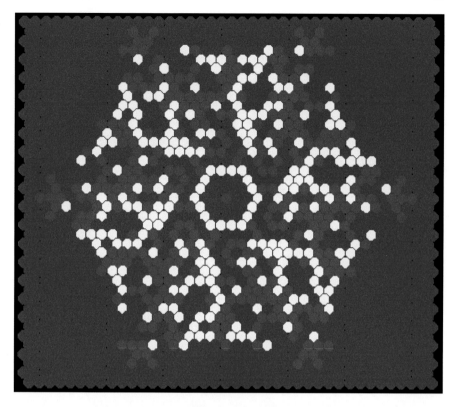

Snowflake image from a 2D isotropic rule. ©2014 Andrew Wuensche.
This is an example of an evolved pattern of a isotropic rule $v=3$, $k=6$, $n=40\times40$, on a 2D hexagonal lattice. Starting from a symmetric pattern such as a single central cell, the symmetry will be conserved by successive time-steps. The actual rule was not recorded so will be forever lost to posterity. However, untold numbers of isotropic rules can be sampled at random to produce analogous snowflake patterns. The image is taken from the DDLab website [166], http://uncomp.uwe.ac.uk/wuensche/multi_value/ddlab_multi_value.html.

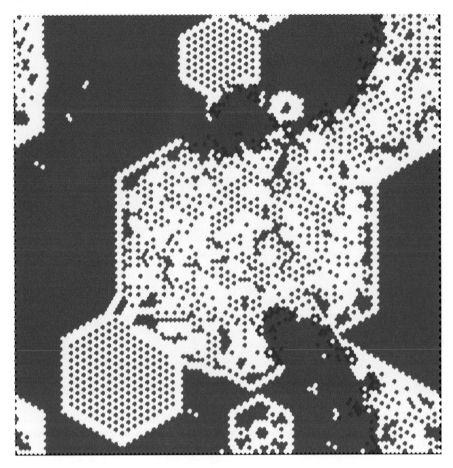

The predator-prey rule. ©2014 Andrew Wuensche.
The predator-prey rule is a $v=3$, $k=6$ k-totalistic rule, kcode=120202210120112111010121110101, n=100x120 on a hexagonal lattice. In k-totalistic rules the lookup table depends just on the totals of each value in the neighborhood. Watching the dynamics unfold, white is able to grow, green feeds on white — otherwise green can barely survive, and red forms a neutral background. The "movie" is more entertaining than one snap-shot. As well as predator-prey, various other interpretations have been made, for example burning paper and forest fire. This rule and similar complex rules were found by an automatic classification method [167, 169]. The image is taken from the DDLab website[166], http://uncomp.uwe.ac.uk/wuensche/multi_value/dd_life.html.

Emergent spinning spirals. ©2014 Andrew Wuensche.
This is a v=3, k=6 k-totalistic rule, kcode=0002000220022001112200021110, n=100×120 on a hexagonal lattice. In k-totalistic rules the lookup table depends just on the totals of each value in the neighborhood. Initially gliders and other structures emerge, but spinning spirals gradually take over the whole space. A version of this rule is shown as a one-value mutation of the "beehive rule" [169] which was found, together with other similar complex rules, by an automatic classification method [167, 169]. The image is taken from the DDLab website [166] http://uncomp.uwe.ac.uk/wuensche/multi_value/dd_life.html.

Spinning spiral boundaries. ©2014 Andrew Wuensche.
The same v=3 (0, 1, 2), k=6 k-totalistic rule and initial pattern as in the previous figure, but about 100 time-steps later, on a 100×120 hexagonal lattice. Here the emerging spinning spirals, which were in a sense competing with each other for territory, have expanded to dominate the whole space. The presentation here is changed to show the boundaries between the competing spirals domains. This is done by arranging 10 frequency bins (2, 4, 6, ..., 20) that collect the frequency of 2s in a moving window of 20 time-steps. The cells are then colored according to the relevant frequency bin. The image is taken from the DDLab website [166], `http://uncomp.uwe.ac.uk/wuensche/multi_value/dd_life.html`, and the method is described in "Exploring Discrete Dynamics"[170].

Larger than Life

Kellie Michele Evans

The images that follow provide a sample of the rich dynamics and complex self-organization exhibited by Larger than Life (LtL), a four-parameter family of two-dimensional cellular automata that generalizes John Horton Conway's celebrated Game of Life (Life) to large neighborhoods and general birth and survival thresholds. LtL was proposed by David Griffeath in the early 1990s to explore whether Life might be a clue to a critical phase point in the threshold-range scaling limit [68]. Extensive experimentation suggests that it is (e.g. [42], [44]). An LtL rule is defined as follows. As with Life, each site of the two-dimensional lattice $\mathbf{Z}^2$ is in one of two states, live or dead. The neighborhood of a site consists of the $(2\rho + 1) \times (2\rho + 1)$ sites in the box surrounding and including it (ρ a natural number). This is called the generalized Moore or "range ρ" box neighborhood. The deterministic update rule is: If a dead site sees between β_1 and β_2 live sites in its range ρ box neighborhood at time t, it will become live at time $t + 1$. Otherwise it will remain dead at time $t + 1$. If a live site sees between δ_1 and δ_2 live sites (including itself) in its neighborhood at time t, it will remain live at time $t + 1$. Otherwise it will become dead at time $t + 1$. Such a rule is written as $(\rho, \beta_1, \beta_2, \delta_1, \delta_2)$. Life has many coherent structures known as still lifes, oscillators, and spaceships [26]. The most intriguing of these structures are the spaceships due to their ability to carry information across long spatial distances. The first three images depict large range versions of Life's spaceships. Extensive experimentation suggests that these LtL structures are quite common, scale in a fairly coherent manner, and have a distinct geometry. The other images depict LtL rules with dynamics vastly different from Life.

K.M. Evans
Department of Mathematics, California State University, Northridge, USA
e-mail: kellie.m.evans@csun.edu

© Springer International Publishing Switzerland 2016
A. Adamatzky and G.J. Martínez (eds.), *Art of Cellular Automata*,
Emergence, Complexity and Computation 20,
DOI: 10.1007/978-3-319-27270-2_3

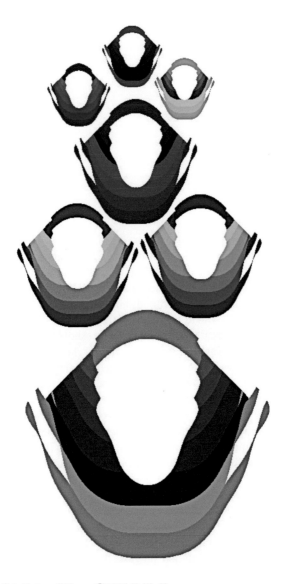

Larger than Life's "winged" bugs ©2015 Kellie Evans.

A Larger than Life *bug* is a coherent structure which generalizes Life's space-ship [43]. The "winged" or "alien" bugs depicted are supported by LtL rules with parameters $(50, 2361, 3120, 2613, 3974)$, $(100, 9350, 12360, 10350, 15740)$, and $(200, 37214, 49194, 41194, 62647)$, respectively, from top to bottom. Each bug was "sculpted" by its respective rule starting from an initial circular region of live site that contains an off-center 'hole' of dead sites. LtL rule parameters and initial circular regions were found using the threshold-range scaling algorithm described in [44]. These winged bugs have the most complex bug geometry seen in LtL rules (to date). The figure is a step toward proving a conjecture from [46]: There is a sequence of winged bugs, each of which is mathematically similar to the next (modulo errors from working in a discrete universe) that converges to a "Euclidean" bug with wings (and a stomach). In the figure, the time at which each site became live is depicted using distinct colors.

Larger than Life's "slow" bugs ©2015 Kellie Evans.
From top to bottom are bugs supported by range 50, 100, and 200 LtL rules, respectively. Each bug was "sculpted" by its respective rule starting from an initial circular region of live site that contains an off-center 'hole' of dead sites. LtL rule parameters and initial circular regions were found using the threshold-range scaling algorithm described in [44]. The specific supporting rule parameters are: $(50, 2773, 4549, 2773, 4700)$, $(100, 10981, 18018, 10981, 18616)$, and $(200, 43706, 71713, 43706, 74093)$. The bugs are "slow," due to not moving very far each time step, as compared to other LtL bugs. This is illustrated in the figure using distinct colors, which represent the time at which a site became live. The "slowness" can be seen in how close together the colored contours are, as compared to, for example, LtL's winged bugs, which have much larger bands of color.

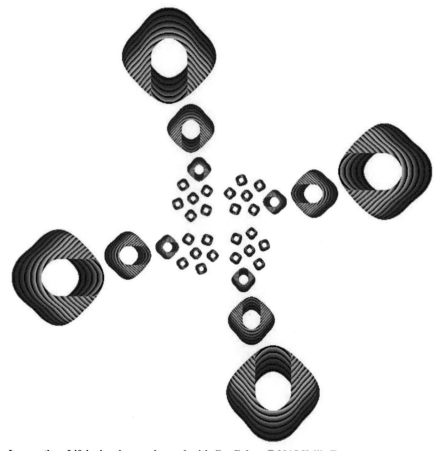

Larger than Life's slow bugs enhanced with GeoGebra ©2015 Kellie Evans.
LtL's range 50, 100, and 200 slow bugs with range 25 slow bugs supported by rule $(25, 706, 1159, 706, 1198)$ in the center. The multi-range configuration of bugs was geometrically transformed using the dynamic mathematics software, GeoGebra http://www.geogebra.org.

Seeing bugs ©2015 Kellie Evans.
Starting from an initial configuration of three bugs supported by a range 50 LtL rule, the rule $(50, 0, 1000, 1, 9200)$ leaves the bugs fixed while generating patterns around them. Color is used to illustrate the time at which sites become live. The figure is shown at time 13, when it has become fixed [47].

Symmetries ©2015 Kellie Evans.
Generated by the LtL rule $(20, 270, 368, 452, 1021)$, starting at time 0 with a circular region of live sites which has radius 99. The grid is size 1200×1200 with wrap around boundary conditions (a.k.a. a torus). See also http://www.csun.edu/~kme52026/invariant.html for a gallery of other such images.

X marks the spot ©2015 Kellie Evans.
The figure shows three copies of an image generated by the LtL rule $(5, 1, 26, 12, 98)$, starting at time 0 with a circle of radius 32 centered at the origin. The grid is size 300×300 with wrap around boundary conditions (a.k.a. a torus). The graphic is shown at time 25 after which it never changes (i.e. it is a still life) [47]. The color variation is due to three different color palettes. See also http://www.csun.edu/~kme52026/invariant.html for a gallery of other such images.

"Maze-like" limiting state ©2015 Kellie Evans.
Starting from product measure with density of live sites below 55 percent or above 59 percent, LtL
rule $(50, 2834, 3975, 2834, 5850)$ either leaves the initial state unchanged or it quickly converges
to all dead sites. However, starting from product measure with density of live sites between
about 56.4 and 58 percent, patches of dead sites appear after 1 time step and then "curves" of
live sites spread outward from the boundary of remaining product measure and also toward its
interior, with "curves" of dead sites between them. Eventually, "maze-like" patterns of live sites
are formed [42]. Color is used to indicate the time at which a live site turns on.

Three Favorite Cellular Automata

Clifford Reiter

Cellular automata (CA) are attractive visually and offer strikingly simple definitions for well motivated models that can produce complex behavior. My study and
research with CA began in earnest the 1990's and since then I have explored many
interesting automata. My research on CA has included models with the symmetry
of snowflakes, self-organizing waves produced by prey-predator models, and cyclic
CA on various nonrectangluar lattices and in three dimensions. In addition, I have
explored many local image processing techniques that are essentially CA for applications to smoothing and enhancing images, and creating Voronoi tilings and other
patterns.

C. Reiter
Lafayette College, Easton, PA, USA
e-mail: reiterc@lafayette.edu

© Springer International Publishing Switzerland 2016   35
A. Adamatzky and G.J. Martínez (eds.), *Art of Cellular Automata*,
Emergence, Complexity and Computation 20,
DOI: 10.1007/978-3-319-27270-2_4

Snow crystal like growth on a real-valued hexagonal lattice. ©2005 Clifford Reiter.
A single cell with value one is immersed in a background of value 0.4. Cells with value greater
than or equal 1 are considered ice. Ice cells and their immediate neighbors called receptive. The
updated value of each cell is the sum of two terms. The first term is the value of the receptive
sites with a constant 0.0001 added; the second term is the average of the non-receptive sites (with
receptive sites contributing values of zero to the averages). Varying the parameters yield different
qualitative behaviors, many of which are snow crystal like [131].

A cellular model for spatial population dynamics self-organizes population migrations
©2010 Clifford Reiter et al.
We update each cell by alternating a local migration scheme (local averaging) with a discrete quadratic prey-predator model. The three populations are shown by red, green and blue intensities respectively. Waves of predators following prey self-organize [39].

Cyclic CA in 3D using Von Neumann neighborhoods ©2011 Clifford Reiter.
Every cell can take one of 38 states. A cell state increases by one (cyclicly) if it has a neighbor with
that state. Otherwise it is unchanged. Four states are shown in the 3D model. After 500 iterations
on a random initial configuration, periodic patterns are partially self-organized [132].

Cellular Automata: Dying to Live Again, Architecture, Art, Design

Robert J. Krawczyk

As in our own lives, outcomes are not predictable. We start in a simple unknown state and through multiple growth and death cycles of individual cells, we become something that could never have been predicted. Can design be explored using a similar process? Can design and art be driven by forces that are unrelated, unattached to preconceptions, unpredictable, and able to explore possibilities not foreseen? What to do you when you don't know what to do? Let the process of life and death take over?

In design, every decision you make sets you off to another course that you could never see, sometimes never imagined. Some decisions spawn affects that are simple and are unrelated to the whole, some are highly integrated and interwoven to other parts in a very complex fashion.

Cellular automata (CA) defines a simple life and death process that has a number of elements that may lend itself to explorations in architecture and art.

Framework: in two-dimensions an equal spacing between nodes or cells in a grid; in three-dimensions the vertical can accommodate an actual height. In three-dimensions, height implies orientation and levels. It implies a base, gravity, and grounding. It also implies that supports should be considered. Cells need to be grounded.

Representation: a cell or a node can be represented as a simple volumetric shape, such as, a square or block, or one that enables cells to form larger contiguous areas and volumes. Such size considerations form large masses or simple overlapping connections. Circles or cylinders can replace squares and blocks. Shapes even become more interesting when their edges are only considered, need not to be solids. In two-dimensions, the overlapping of shapes becomes a linking mechanism so cells are not orphaned in space.

R.J. Krawczyk

Illinois Institute of Technology, College of Architecture, 3360 South State Street, Chicago, IL 60616, USA

e-mail: krawczyk@iit.edu

© Springer International Publishing Switzerland 2016

A. Adamatzky and G.J. Martínez (eds.), *Art of Cellular Automata*,
Emergence, Complexity and Computation 20,
DOI: 10.1007/978-3-319-27270-2_5

Transitions: simple rules based on neighbourhood relationship are used to determine who lives, who dies, and who stays to propagate. Transitional rules can include a variety of neighbours depending if the framework is two or three dimensional. Transitions are the basis for generations. Transitions can be quite repetitive or can be mutated to introduce the unanticipated; unexpected changes. These mutations can be triggered randomly.

Time: the present is always being created and past can be captured. Ability to remember past generations offers the envisioning of an entire life span of the cells. The present determines the future; every cell plays a role in creating the next generation. Enabling the capture of history offers glimpses into the past and a foundation for the present to build on. The number of generations determines the total expanse, the universe, density of cells, infinitely repeating patterns or possibly extinction.

Instigation: a starting state, an initial arrangement of cells is necessary. Is it purposeful, does it reference something, is it random, is it symmetrical, or non-symmetrical? All these can have a different meaning if the framework is two or three dimensional.

All of these elements form the rich variety that begins to address any exploration. They enable the possibly of surprise, the unexpected, or even chaos. They comprise the antithesis of parametric design; know your boundaries, pick your parameters, and run them through tightly confined, safety selected iterations. Here natural chaos rules.

To repeat: what do you do when you don't know what to do; you let it happen on its own terms, stand back and enjoy the results.

The following examples attempt to explore some of these ideas: massing, connectivity, linking, two-dimensions, three-dimensions; resulting in drawings, images, rendered models, and 3D prints.

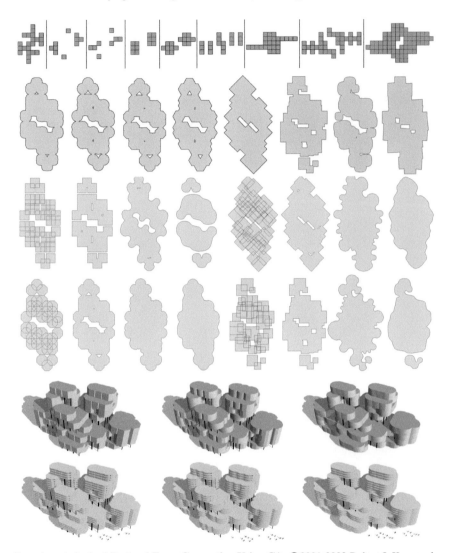

Experiments in Architectural Form Generation Using CA. ©2001-2002 Robert J. Krawczyk. These experiments extend CA to three dimensions and to suggest architectural forms; first, by including explicit architectural considerations, such as, individual space units, a supporting structure, and the development of an envelope, and secondly, by retaining growth as the cells survive generations. The retention of growth allows the forms to explode from the underlying lattice it started in. This image displays for evolving CA, variety of 3D module shapes, some overlapping others, the development of edge of each level, and finally the 3D form as a mass model and one with levels [84]. See also images at: http://iit.edu/~krawczyk/acsa01/acsa01.htm

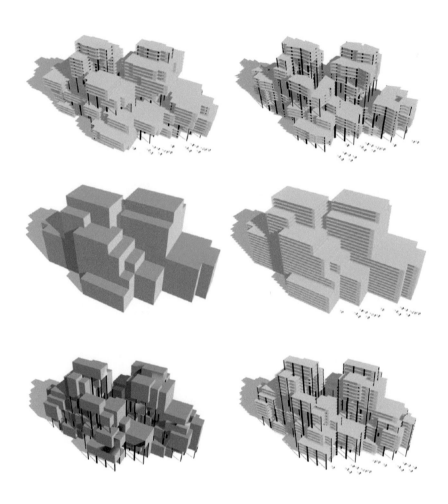

Experiments in Architectural Form Generation Using CA ©2001-2002 Robert J. Krawczyk.
Additional examples of the development of possible 3D massing models, with ones where
levels and structure are indicated [83]. See also images at: http://iit.edu/~krawczyk/
acsa01/acsa01.htm

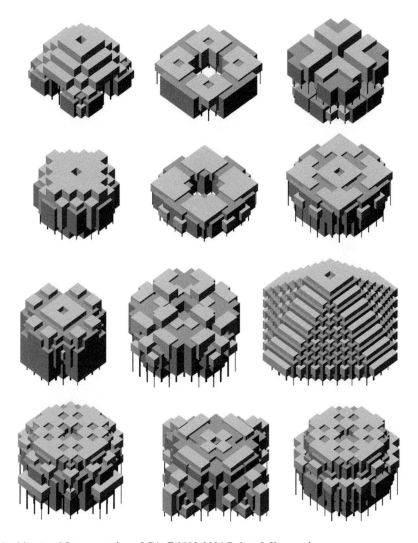

Architectural Interpretation of CA ©2002-2004 Robert J. Krawczyk.
As a cell survives generation after generation, its size is increased to be able to develop a mass that better embodies the time element within the cellular automata process. Architecturally, the process develops connecting masses which further enhance and highlight the method used to create them. The standard method treats all cells equally. In this series, every combination of boundary, rule, and neighbourhood is generated for a set of life spans. This includes two boundary conditions, one limited and the other unlimited; thirty-seven rules, based on a survival/birth neighbourhood count, and four neighbourhood definitions; each pass through four, five, six, nine, and fifteen generations. The initial configuration consists of eight cells in a square arrangement having the centre cell empty. The initial configuration consists of eight cells in a square arrangement having the centre cell empty. The cells are represented as rectangular volumes [83]. See also: http://iit.edu/~krawczyk/nks2003

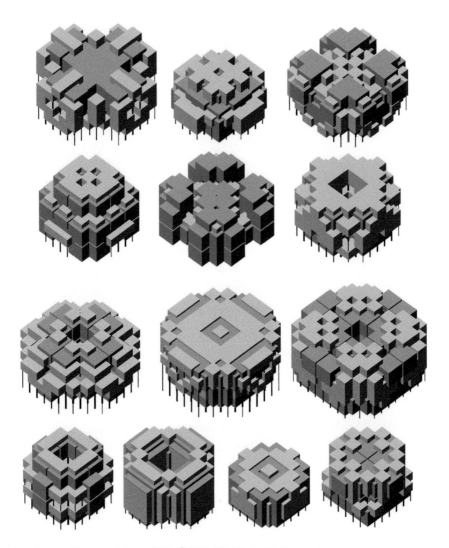

Architectural Interpretation of CA ©2002-2004 Robert J. Krawczyk.
As the cells pass from one generation to another they are normally subjected to the same survival/birth rule based on the same neighbourhood count. This experiment explores the concept that at each generation a mutation is applied by randomly selecting a new rule and neighbourhood count. Architecturally, the concept explores a method to break any evolving pattern so the forms are further unpredictable and offer an even wider range of configurations without introducing a natural style The boundary is set to unlimited throughout with the random selection using all thirty-seven rules and all four neighbourhood types. In this series the life span is set to six generations and the space module is represented as a cube. The initial configuration consists of eight cells in a square arrangement having the centre cell empty. The cells are represented as rectangular volumes [85]. See also: http://iit.edu/~krawczyk/nks2003

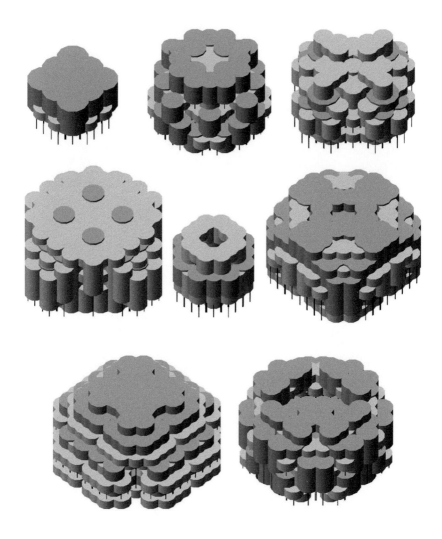

Architectural Interpretation of CA ©2002-2004 Robert J. Krawczyk.
As the cells pass from one generation to another they are normally subjected to the same survival/birth rule based on the same neighbourhood count. This experiment explores the concept that at each generation a mutation is applied by randomly selecting a new rule and neighbourhood count. Architecturally, the concept explores a method to break any evolving pattern so the forms are further unpredictable and offer an even wider range of configurations without introducing a natural style. The boundary is set to unlimited throughout with the random selection using all thirty-seven rules and all four neighbourhood types. In this series the life span is set to seven generations and the space module is represented as a cylinder. The initial configuration consists of eight cells in a square arrangement having the centre cell empty. The cells are represented as cylindrical volumes [85]. See also: http://iit.edu/~krawczyk/nks2003

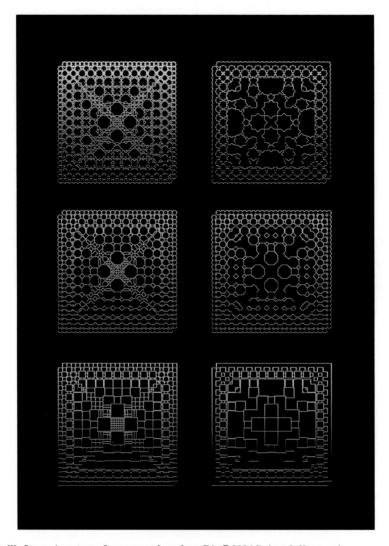

Metallic Lace: A system of ornament based on CA ©2006 Robert J. Krawczyk.
These images panels show a set of CA variations exploring the interpretation of history in
the simulation based on a cellular automata algorithm. The history of each cell is recorded by
accumulating the number of times a cell is visited and then this number is used as a size factor
for each cell. Finally the cells are made solid and then the edges are rendered. Twenty-seven
different patterns were fully investigated from the initial generation of 450; each in 12 variations.
The cell types were translated as is, or joined to form larger areas. Architecturally these are
seen as bas relief patterns, wrought iron ornamentation, or sandblasted or laser etched glass
patterns. Entry prepared for the New Kind of Science 2006 Conference Art Gallery. See also:
http://bitartworks.com/lace/index.html

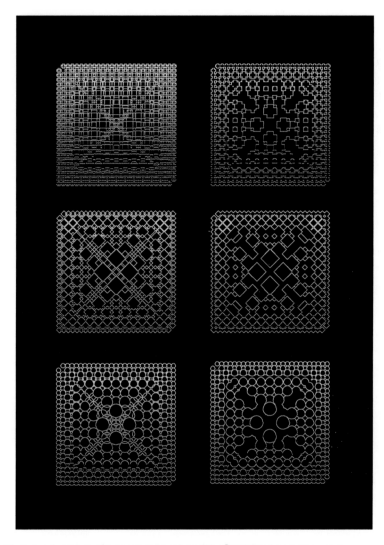

Metallic Lace: A system of ornament based on CA. ©2006 Robert J. Krawczyk.
During this exploration the challenge was to graphically demonstrate the cell history. The
variations shown include a single 12 generation pattern in a variety of cell types: circle,
square, diamond, octagon, a cross, and an 8-point star. These were all attempts to trans-
late the raw cellular automata results into an architectural design element. The original
lines that created the pattern were rendered as "metallic lace" to give them a third dimension. Entry prepared for the New Kind of Science 2006 Conference Art Gallery. See also:
http://bitartworks.com/lace/index.html

Metallic Lace to Ring Cross: 2D Sketches ©2001-2002 Robert J. Krawczyk.
The exploration of CA to generate two-dimensional ornamentation is extended into three-dimensions. The pattern is based on generations of survival, birth, and dead. As the pattern evolves, for each cell that is occupied, its location and the number of times it is visited is recorded. This measure of history is then used to determine the rings three-dimensional location and its size; forming multiple layers. The cells for these initial 2D sketches included: circle, square, diamond, octagon, a cross, and an 8-point star. The 2D version of these could be used to filter light if the patterns were sand blasted in glass. Much of the same methods of the Metallic Lace were incorporated here. See also: http://bitartworks.com/lace/lace01/gallery02.html

Ring Cross: 3D Model Sketches. ©2005 Robert J. Krawczyk.
This measure of history is then used to determine the rings three-dimensional location and its size. This ornamentation can be a 3D grille to filter for light through or part of a building elevation as a false front. It could even be used for framing glass in a clear-story window or a section of overhanging shading. The final versions used the circle cell type.

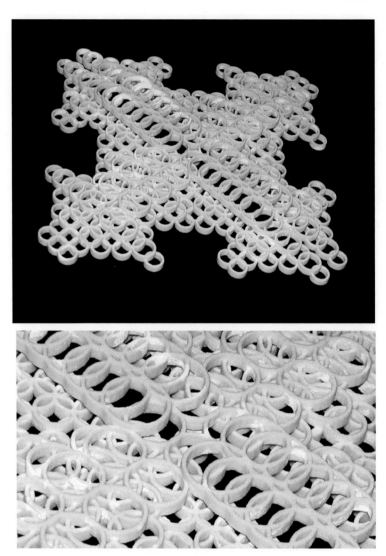

Ring Cross 17140112: 3D Printed Prototype ©2005 Robert J. Krawczyk.
This prototype was produced using the Z-Corp 3D Printer, produced in plaster; measures 10"×10" with five interlocking layers of a circular module. It was entered in the 4th Digital Sculpture Competition, 2005, Ars Mathématica, Paris France.

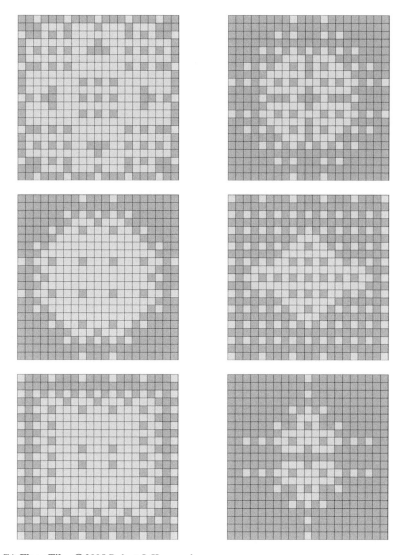

2D CA Floor Tiles ©2005 Robert J. Krawczyk.
The initial sketches for a tile design are based on a cellular automata algorithm, the scale being 21×21 set of tiles in a 10×10 meter area. The general concept is to use a very large number of unique tile designs throughout the campus. Each different area could be identified by its own family of designs. The initial sketches were based on tiles of two complimentary earthy colours. This image displays a catalog of designs that have symmetry across either the horizontal or vertical axis. Initial states, generational rules, number of generations, and neighbourhood criteria were randomly selected. Over 3,000 designs were generated for the tiles.

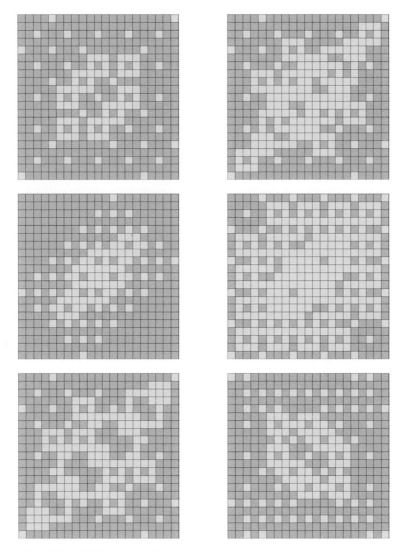

2D CA Floor Tiles ©2005 Robert J. Krawczyk.
The initial sketches for a tile design are based on a CA algorithm, the scale being 21×21 set of tiles in a 10×10 meter area. The general concept is to use a very large number of unique tile designs throughout the campus. Each different area could be identified by its own family of designs. The initial sketches were based on tiles of two complimentary earthy colours. This image displays a catalog of designs that have some symmetry across the diagonal. Initial states, generational rules, number of generations, and neighbourhood criteria were randomly selected. Over 3,000 designs were generated for the tiles.

In Search of Movement and Life on a Static Grid

Tim J. Hutton

The search for life-like processes that can be simulated on computer has led to many types of proposed systems. Of these, cellular automata are one of the most venerable and one of the most visually appealing.

There are many software packages for experimenting with cellular automata. Our images are produced by two of them, both free and open source. The first is Golly [135], created by Andrew Trevorrow and Tom Rokicki. Golly uses Bill Gosper's hashlife algorithm which exploits spatial and temporal repetitions to run many patterns at extraordinary speed.

The second piece of software we have used is Ready [76] which is aimed more at continuous-state cellular automata and supports different meshes. Ready uses OpenCL which makes the GPU do all the hard work, running many of the update steps in parallel.

As community projects, both Golly and Ready make it easy for users to share their patterns, and provide a wide variety of systems with astonishing behavior. Some of our systems are chosen for explicitly aesthetic reasons while others have a functional aesthetic, visible only when the viewer has penetrated their workings sufficiently to be astounded that such a behavior is even possible.

T.J. Hutton
e-mail: tim.hutton@gmail.com

© Springer International Publishing Switzerland 2016   53
A. Adamatzky and G.J. Martínez (eds.), *Art of Cellular Automata*,
Emergence, Complexity and Computation 20,
DOI: 10.1007/978-3-319-27270-2_6

The 'Computer Art' Turmite ©2015 Tim Hutton.

A turmite is an extension of Langton's Ant [86] which walks on a grid and can change the color of its square and turn left or right. This turmite has two states and two colors and was discovered by Ed Pegg, Jr. It behaves as follows: When in state 0 on a black square, it changes the square to white, turns left, moves forwards one square and stays in state 0. When in state 0 on a white square, it leaves it white, turns right, moves forwards one square and enters state 1. When in state 1 on a black square, it leaves it black, turns right, moves forwards one square and enters state 0. When in state 1 on a white square, it changes the square to black, turns left, moves forwards one square and stays in state 1. Starting in state 0 on an empty black grid, after 1,217,124 steps this turmite produces the picture shown, making its own art in a frame that expands. This system was run as an 18-state cellular automaton in Golly [135].

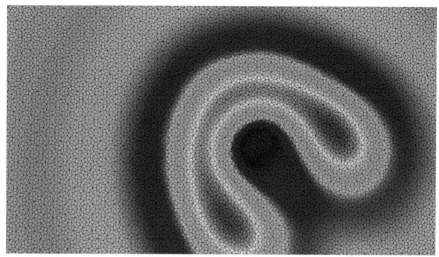

The Munafo U-Skate glider running on a Penrose P3 aperiodic tiling. ©2015 Tim Hutton.
Every cell in the tiling has two continuous values, u and v, with their rates of change given by the
Gray-Scott reaction-diffusion system: $\frac{du}{dt} = D_u \nabla^2 u - uv^2 + F(1-u)$, $\frac{dv}{dt} = D_v \nabla^2 v + uv^2 - (F+k)v$.
Robert Munafo made the discovery that when $F = 0.062$ and $k = 0.06093$ it is possible to have
stable moving patterns - gliders. The one shown moves up and to the left. Despite the irregularity of
the underlying grid the glider manages to maintain its shape and continue on its way. For a moving
version of this, see the video at [75]. The diffusion coefficients are $D_u = 1.312, D_v = 0.656$ and the
system is simulated with explicit Euler integration, with timesteps of $\Delta t = 0.2$. ∇^2 is the Laplacian
operator. Image produced with Ready [76]. Visit Robert's page at [125].

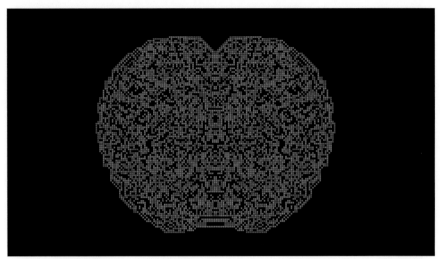

Langton's Ant LLRR ©2015 Tim Hutton.
This image shows the output of an extension of Langton's Ant [86] which walks on a grid and can change the color of its square and turn left or right. This one draws with four colors, always replacing one with the next in the list in a cyclic fashion. It also turns on each color: left, left, right and right, respectively, hence it is known as the LLRR ant. Starting on an empty black grid, after 2,590,355 steps this ant produces the picture shown. Other similar systems are asymmetrical but this one displays an intriguing symmetry in a cardoid shape that persists as far as is known. There is an interesting proof that this ant will continue to produce a symmetrical shape, which involves Truchet tiles [58]. This system was run as a 20-state cellular automaton in Golly [135].

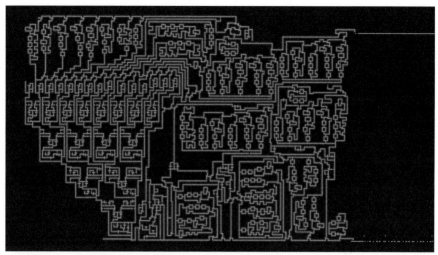

John Devore's self-replicating computer. ©2015 Tim Hutton.
In the 1940's, John von Neumann had shown how to make a self-replicating computer in a 29-state CA [163]. In 1968, Edgar Codd had shown (reportedly to win a bet in a pub [66]) that it was possible to do it with only 8 states, but it required a design of millions of cells [29, 77]. In 1973, John Devore was able to reduce Codd's machine by several orders of magnitude in this elegant design [36]. There's a program tape extending from the bottom-right, and a data tape in the top-right. The construction arm extends from the top-left of the machine, and after copying the whole pattern can inject a 'go' signal into a port on the bottom left of the new copy to set it running. This image was produced by Golly [135], which can run the pattern for the 1.02×10^{11} steps needed to see the replication happen. When designing his machine in the early 70's, John worked on an IBM 360/50 computer, without a monitor — printing the pattern out every 1000 to 5000 steps. He told me, "I still have a stack of printouts that have never been trimmed and taped."

Matt Pennybacker's single chemical model. ©2015 Tim Hutton.
Alan Turing created reaction-diffusion systems in the 1950's [149], presenting a system with two
chemicals — an 'activator' and an 'inhibitor'. And since then reaction-diffusion systems such as
Gray-Scott and the Brusselator (a model of the famous Belousov-Zhabotinsky reaction) have also
used at least two chemicals. This model however, suggested by Matt Pennybacker [129], uses
only a single value at each cell, and still manages to produce the spots and stripes phenomena
that characterize reaction-diffusion systems such as Turing's. The rate of change is given by $\frac{du}{dt} =
2u - \nabla^4 u - 2\nabla^2 u - u^3$. ∇^2 is the Laplacian operator and ∇^4 is the bilaplacian — the Laplacian
applied twice. So if you compute your Laplacian with a 3×3 pattern of neighborhood weights
[0,1,0; 1,-4,1; 0,1,0] then the weights for the bilaplacian are a 5×5 pattern: [0,0,1,0,0; 0,2,-8,2,0;
1,-8,20,-8,1; 0,2,-8,2,0; 0,0,1,0,0].

Some Beautiful and Difficult Questions about Cellular Automata

Nathaniel Johnston

One of the most interesting things about cellular automata (CA) is that such simple sets of rules can give rise not only to such strange and unpredictable behavior, but also that it can create *seemingly* predictable behavior that is nonetheless extremely difficult to actually pin down. One example of this apparently simple behavior comes from the "Life without Death" cellular automaton, which is the exact same as Conway's Game of Life with the exception that cells never die. Just by playing around with random starting patterns, it immediately becomes "clear" that the majority of sufficiently large patterns explode quadratically in this rule and fill most of the 2D plane. However, it was not even known if there is a *single* quadratically-growing pattern in this rule until Dean Hickerson found one in 2009, over 20 years after this CA was first studied.

The "toothpick"-type cellular automata are another family of CA that seem quite regular and predictable at first glance, but nonetheless hold some interesting mathematics under the hood. The general idea with these CA is to start by drawing some shape (such as a line segment), then at each generation draw more copies of that shape at certain parts of the shapes that are already present (possibly rotated in some pre-specified way). For example, the original toothpick CA [20] starts with a single line segment (a "toothpick") drawn horizontally. Then toothpicks are drawn vertically at the endpoints of the original toothpick, creating an "H" shape. Then toothpicks are drawn horizontally at the endpoints of the vertical toothpicks, and so on, back and forth. After running this CA for a while, it becomes "clear" that it expands extremely quickly until a square gets filled in, then becomes quite slow and gradually speeds up until a larger square gets filled in, and then this process repeats. However, actually characterizing exactly how quickly it grows in between the two extremes (right before and right after each square is filled in) leads to some surprisingly deep mathematics. The "Q-toothpick" CA pictured below behaves quite similarly, but is based on drawing quarter circles rather than straight lines.

N. Johnston
Mount Allison University, New Brunswick, Canada
e-mail: njohnston@mta.ca

© Springer International Publishing Switzerland 2016
A. Adamatzky and G.J. Martínez (eds.), *Art of Cellular Automata*,
Emergence, Complexity and Computation 20,
DOI: 10.1007/978-3-319-27270-2_7

A more well-known question comes from Conway's Game of Life itself, which has long been known to have "Gardens of Eden": patterns that no other pattern evolves into. The first explicit Garden of Eden ever constructed in the Game of Life was found in 1971, but ever since then people have wanted to know just how small they can be. Progress on this problem has been made very steadily over the years, with record-breaking Gardens of Eden being found in each of 1991, 2004, 2009, and 2011. However, there is still a long way to go: it is known that no Garden of Eden fits within a 6-by-6 square, but the smallest known Garden of Eden takes up most of a 10-by-10 square.

Quadratic growth in the "Life without Death" automaton. ©2015 Nathaniel Johnston
which behaves exactly the same as Conway's Game of Life, except that cells can never die. This
CA is interesting for the fact that it naturally sprouts "ladders" that can be used to simulate arbitrary
boolean circuits [67]. Despite the fact that most randomly-generated starting patterns of sufficient
size seem to explode and grow quadratically, none were explicitly proved to have this property [64]
until Dean Hickerson found this configuration in 2009.

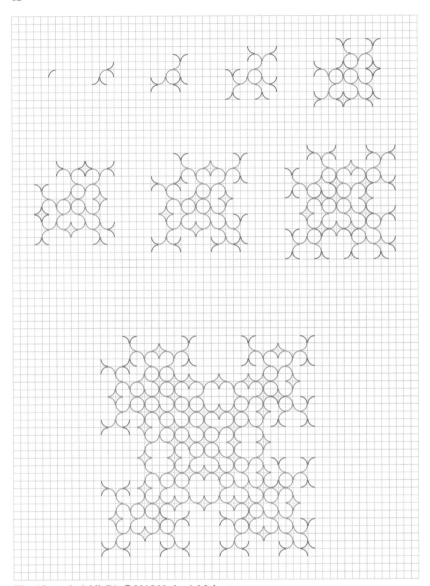

The "Q-toothpick" CA.©2015 Nathaniel Johnston
in which any exposed endpoint of a quarter circle in one generation sprouts two new quarter circles in the adjacent cells in the next generation. The figure demonstrates generations 1 – 8 and generation 15 of the evolution of a single quarter circle in this automaton. Newly-added quarter circles are red, while pre-existing quarter circles are green. This CA fills the plane with a pattern consisting of circles, diamonds, hearts, and other bulbous shapes. It was inspired by, and behaves very similar to, the "toothpick" automaton that was studied in [20].

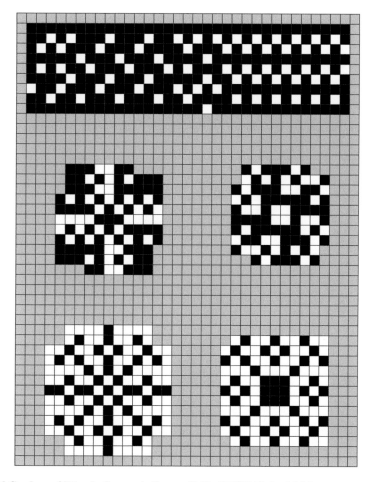

Several Gardens of Eden in Conway's Game of Life, ©2015 Nathaniel Johnston
with black cells being alive, white cells being dead, and the state of grey cells being irrelevant
(they can be alive or dead). That is, these are patterns which cannot be evolved into, no matter
what the initial configuration of the Life grid is. The first Garden of Eden to ever be found (top)
was published by Roger Banks, Mike Beeler, Steve Ward, and Rich Schroeppel in 1971, and since
then there has been a push to find the smallest Garden of Eden that exists. Recent searches have
focused on trying to find Gardens of Eden with various forms of symmetry, in order to reduce
the size of the search space, and the resulting patterns are quite beautiful. The "Flower of Eden"
(middle left) was found by Nicolay Beluchenko in 2009, and was the smallest known Garden of
Eden until the three other Gardens of Eden above were found in [71]. The middle-right Garden of
Eden is the smallest known by the number of total cells that must be specified (92), the bottom-left
one has the lowest known density (0.320), and the bottom-right one has the lowest known number
of alive cells (45). An exhaustive computer search was also carried out in [71] that showed that no
Garden of Eden fits within a 6-by-6 square.

Hyperbolic Gallery

Maurice Margenstern

We give five illustrations of cellular automata (CA) in hyperbolic spaces. The text
under each picture indicates the paper from which it is taken together with some
indications on the meaning of the picture. In order to help the reader to grasp something about what the meaning of each picture is, here we give a short introduction
to hyperbolic spaces and then the general context of my research. The below figure gives a schematic representation of the hyperbolic plane in what is called the
Poincaré's disc.

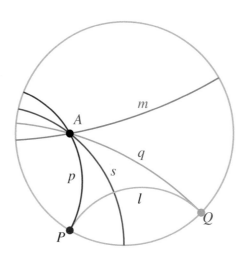

Illustrating Poincaré's model of the hyperbolic plane.

M. Margenstern
Université de Lorraine, Metz, France
e-mail: margenstern@gmail.com

© Springer International Publishing Switzerland 2016
A. Adamatzky and G.J. Martínez (eds.), *Art of Cellular Automata,*
Emergence, Complexity and Computation 20,
DOI: 10.1007/978-3-319-27270-2_8

In this figure, the points which are inside the disc, say D, represent the hyperbolic plane. The lines of the plane are the trace in D of diameters or of circles which are orthogonal to ∂D, the circle which constitutes the border of D. The figure shows us a point A which is outside a line ℓ. The lines s, p, q and m pass through A. The line s cuts ℓ in the hyperbolic plane. The lines p and q are both parallel to ℓ: each one shares a point of ∂D with ℓ. Such a point, called point at infinity does not belong to the hyperbolic plane. At last, the line m does not cut ℓ: neither in the hyperbolic plane, nor on ∂D, nor outside D. This environment can be extended to any dimension n: replace D by a nD-hyper-ball, say, B, and replace ∂D by the border ∂B of B. We say that ∂B is an nD-hypersphere. Hyperbolic hyperplanes are diametral hyperplanes or trace in B of hyper-spheres which are orthogonal to ∂B. We leave the other details to the reader which may have a look at [93].

In the hyperbolic plane, there are infinitely many different tessellations which are generated by a regular convex polygon P. Let p be the number of sides of P. Then, the interior angle at each vertex of P is of the form $\dfrac{2\pi}{q}$, where q is a positive integer, at least 3. A corollary of Poincaré's Theorem, see [93], says that if p and q satisfy the inequality $\dfrac{1}{p} + \dfrac{1}{q} < \dfrac{1}{2}$, then P tiles the hyperbolic plane. Such a tessellation is denoted by $\{p,q\}$. Our illustrations will take place in $\{5,4\}$, the pentagrid and also in $\{7,3\}$, the heptagrid. We shall also meet the tessellation $\{5,3,4\}$ of the hyperbolic 3D-space, called the dodecagrid. It is built on Poincaré's dodecahedron: it has 12 faces which are rectangular regular pentagons. Three of them share a vertex of the dodecahedron and there are four such dodecahedra around an edge of each dodecahedron of the tessellation.

My research is mainly concerned with CA in hyperbolic spaces. The illustrations which I give further take place in this context. Three of them deal with universal CA with as few states as possible. As indicated later, universal CA, say UCA, means different objects. Later, we shall distinguish strongly UCA and weakly UCA. A strongly universal CA directly or indirectly simulates a Turing machine from a finite configuration: this means that all cells are initially blank expected finitely many of them. Also, the simulating UCA stops its computation if and only if the simulated machine halts, *i.e.* it stops its own computation on the given data. In a weakly UCA, either one of these constraints or both of them are relaxed. The initial configuration may be infinite but not arbitrarily infinite: it must be periodic at large. Also, the simulating automaton may not stop when the simulated Turing machine halts. In that case, it signalises the halting.

The first figure deals with another tiling which contains the heptagrid as a sub-tiling. The last figure is a sub-tiling of the heptagrid which deals with the tiling problem: is it possible to find an algorithm which, given the description of a finite set of tiles, says whether or not it is possible to tile the hyperbolic plane with copies of these tiles.

A CA in a grid of the hyperbolic plane which evolves like colonies of bacteria. ©2015 Maurice Margenstern.

The tiling is obtained from the heptagrid in two steps: first by dividing each heptagon into seven equal ones from the centre of the heptagon; then, each such new tile T is divided into four triangles whose vertices are taken among the mid-points of the sides of T. The figure is similar to a picture of [95]. The picture of that paper belongs to a series of 36 pictures starting from time 0 up to time 35. The picture represents the configuration reached by the cellular automaton at time 35, the configuration at time 0 being reduced to the central blue heptagon. The series of 36 pictures evolves, in time, like a colony of bacteria in low nutrient conditions.

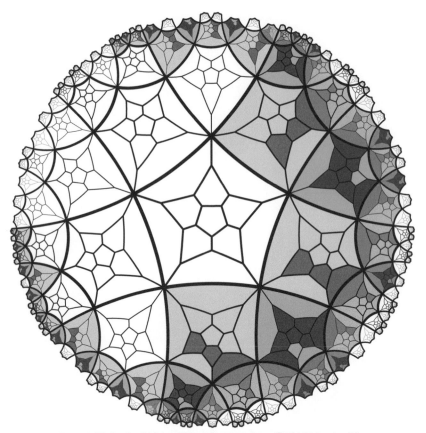

Weakly universal CA in the dodecagrid with two states. ©2015 Maurice Margenstern.
The automaton of this picture as well as that of the next one are weakly universal cellular
automata. In both cases, they simulate the motion of a single 'locomotive' in a 'railway circuit'.
Such a model is able to simulate any Turing machine, see [148]. For both automata, the railway
circuit implements any register machine with two registers: it is known that this latter model
is universal, see [121]. The automaton illustrated by this picture appeared in [112]. The figure
shows the implementation of what is called a horizontal track in [112]. As underlined in that latter
paper, the implementation of the tracks is a non-trivial task in the proof. It is also a mandatory
one: the simulation of gates ensures logical completeness only which is very far from Turing
completeness. In the implementation illustrated by this picture, tracks are one way. Both ways are
implemented along the same pattern: yellow and green cells for one direction, orange and purple
cells for the other. Note that the colours do not represent the states of the automaton. A light
colour corresponds to the blank. A dark colour corresponds to the other state of the automaton.

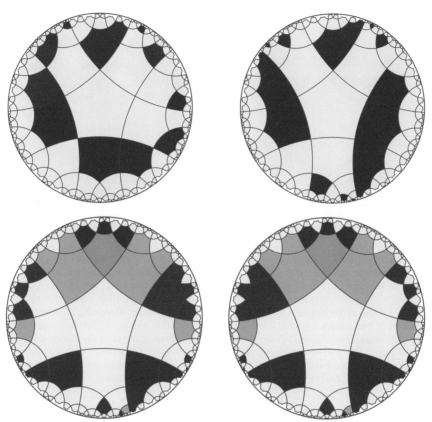

Smallest weakly universal CA in the pentagrid. ©2015 Maurice Margenstern.
It is the smallest weakly universal CA in the pentagrid with respect to the number of states: here, five of them. In the heptagrid, it was possible to lower the number of states down to three of them. The CA illustrated by the picture, as well as its 3D relatives, and those which live in the heptagrid are rotation invariant. These automata are non-linear ones in this meaning that the circuit contains infinitely many cycles of cells which are not in the quiescent state. The figure shows the idle configurations of the crossings and two switches which are parts of the nodes of the circuit constituted by the tracks and by the locomotive. Here, the tracks are one way. This induces a double structure for the memory switch, one called active and the other passive [97]. The pictures illustrate idle configurations. Above, from left to right: the crossing, and the fixed switch. Below, the two versions of the flip-flop: to the left-hand side track and to the right-hand side one.

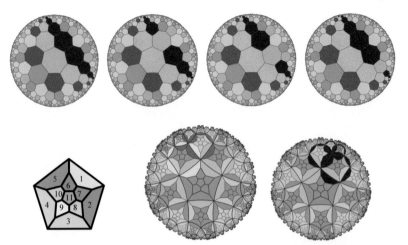

Strongly universal CA in the heptagrid and in the dodecagrid. ©2015 Maurice Margenstern. These automata are constructed in [96]. They are strongly universal CA on the line, in the penta-grid, in the heptagrid and in the dodecagrid. This time, the CA does not mimic the behaviour of a locomotive on a railway circuit. It implements a CA on the line which is almost strongly uni-versal. First line of the picture: in the heptagrid, a series of pictures illustrating the detection of the halting of the simulated machine, which triggers the process which will stop the computation of the automaton. Second line: the basic patterns for implementing the same one dimensional CA in the dodecagrid. The leftmost drawing shows the numbering of the faces of a dodachedron. The middle picture shows the cells which are in the upper half-space defined by the basic plane of the simulation. The rightmost picture illustrates the half-space below that plane.

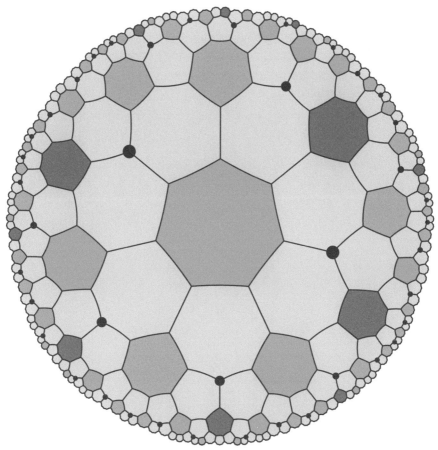

The mantilla: a sub-tiling of the heptagrid, a tessellation of the hyperbolic plane. ©2015 Maurice Margenstern.

The result of the construction after a huge number of iterations of the automaton. The picture is adapted from the representations of the mantilla given in [94]. It is a sub-tiling of the heptagrid, the tessellation $\{7,3\}$, which can recursively be generated by three rules [94]. From these rules, it is not difficult to devise a cellular automaton which will construct the tessellation starting from a tile, of course with an infinite time. Tessellations are closely related with cellular automata in the plane or in the 3D-space: in the standard Euclidean setting, this is the case. Cells are most often represented by squares, sometimes by regular hexagons, in a few cases by equilateral triangles. As reminded the reader in our introduction, there are infinitely many tessellations in the hyperbolic plane. The mantilla was used in [94] to show that there is no algorithm which would solve the tiling problem in the hyperbolic plane.

Evolved Gliders and Waves on a Geodesic Grid

Jeffrey Ventrella

Cellular automata (CA) are typically modeled on 1D, 2D, or 3D grids. To avoid boundary artifacts, it is common to create periodic boundary conditions so that opposing boundaries wrap around. In the case of a 2D cellular automaton, this periodic boundary creates the topological equivalent of a torus. A 2D grid can be mapped onto a 3D torus without introducing cell neighborhood discontinuities. This mapping introduces some geometrical distortion, but no topological artifacts — no discontinuities in grid connectivity. This is not the case when mapping a grid onto a sphere, which has positive curvature everywhere.

For any two points on a sphere, if they are each set to motion and follow geodesic paths, they will come back to their original starting points. Interestingly: the two paths will intersect twice (unless the paths are identical). This is due to the fact that all lines drawn on a sphere are actually geodesic arcs that close to form great circles. All great circles intersect all other great circles. How might this fact about spherical geodesics affect CA dynamics where glider and wave collisions are important?

Any implementation of a CA requires some kind of grid of cells. A grid can be drawn on a sphere — but there will inevitably be 'pinch-points' (specific locations in the grid where vertex valence is smaller than average). When progressively triangulating a sphere — as in the process to make a geodesic dome these pinch-points correspond to the vertices of a platonic solid. For the images shown here, the icosahedron was used. Given a CA rule that supports gliders with complex interactions, as in Wolfram's Class IV CA dynamics, and given the discontinuities that result from mapping a regular grid onto a sphere, what new and different dynamics can emerge?

To identify a uniquely spherical type of glider dynamics, a variation of an XOR gate was implemented using the Game of Life rules on a square grid mapped to a sphere (placing a rectangular grid on each face of a cube and puffing them out to the shape of a sphere). This XOR gate avoids the pinch points and takes advantage of the extra crossing that occurs from the positive spherical curvature [158].

J. Ventrella
10 4th St. Petaluma, CA, 94952, USA
e-mail: jeffreyventrella@gmail.com

© Springer International Publishing Switzerland 2016   73
A. Adamatzky and G.J. Martínez (eds.), *Art of Cellular Automata*,
Emergence, Complexity and Computation 20,
DOI: 10.1007/978-3-319-27270-2_9

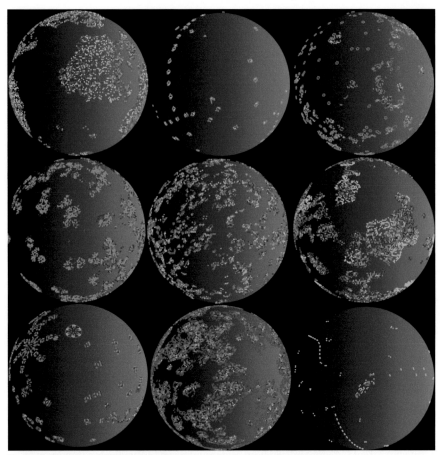

Evolved Gliders and Waves on a Geodesic Grid ©2015 Jeffrey Ventrella.
A geodesic grid with icosahedral symmetry determines cell positions and neighbor topology (all cells have 6 neighbors except for 12, which have 5). Each cell can have 8 possible states (shown in Earth colors). The transition rule takes into account the number of neighbors for multiple states using a set of sub-rules that are applied in sequence at time t to determine the final state at time $t + 1$. Transition rules are evolved using an interactive tool for breeding gliders. The technique is described in [159]. The web site http://www.ventrella.com/earthday/ allows for viewing these CA, and the web site http://www.ventrella.com/BreedingGliders/ demonstrates interactive breeding of gliders.

Constructing Counters through Evolution

Moshe Sipper

In the one-dimensional synchronisation task, discussed in [147], the final pattern consists of an oscillation between all 0s and all 1s. From an engineering point of view, this period-2 cycle may be considered a 1-bit counter. Building upon such an evolved CA, using a small number of different cellular clock rates, 2- and 3-bit counters can be constructed.

Constructing a 2-bit counter from a non-uniform, radius $r = 1$ CA, evolved to solve the synchronisation task, is carried out by "interlacing" two $r = 1$ CAs, in the following manner: each cell in the evolved $r = 1$ CA is transformed into an $r = 2$ cell, two duplicates of which are juxtaposed (the resulting grid's size is thus doubled). This transformation is carried out by "blowing up" the $r = 1$ rule table into an $r = 2$ one, creating from each of the (eight) $r = 1$ table entries four $r = 2$ table entries, resulting in the 32-bit $r = 2$ rule table. For example, entry $110 \rightarrow 1$ specifies a next-state bit of 1 for an $r = 1$ neighbourhood of 110 (left cell is in state 1, central cell is in state 1, right cell is in state 0). Transforming it into an $r = 2$ table entry is carried out by "moving" the adjacent, distance-1 cells to a distance of 2, i.e., $110 \rightarrow 1$ becomes $1X1Y0 \rightarrow 1$; filling in the four permutations of (X, Y), namely, $(0,0)$, $(0,1)$, $(1,0)$, and $(1,1)$, results in the four $r = 2$ table entries. The clocks of the odd-numbered cells function twice as fast as those of the even-numbered cells, meaning that the latter update their states every second time step with respect to the former. The resulting CA converges to a period-4 cycle upon presentation of a random initial configuration, a behaviour that may be considered a 2-bit counter.

Constructing a 3-bit counter from a non-uniform, $r = 1$ CA is carried out in a similar manner, by "interlacing" three radius $r = 1$ CAs (the resulting grid's size is thus tripled). The clocks of cells $0, 3, 6, \ldots$ function normally, those of cells $1, 4, 7, \ldots$ are divided by two (i.e., these cells change state every second time step with respect to the "fast" cells), and the clocks of cells $2, 5, 8, \ldots$ are divided by four (i.e., these cells change state every fourth time step with respect to the fast cells). The resulting CA converges to a period-8 cycle upon presentation of a random initial configuration, a behaviour that may be considered a 3-bit counter. We have thus demonstrated how one can build upon an evolved behaviour in order to construct devices of interest.

M. Sipper
Ben-Gurion University, Beer Sheva, Israel
e-mail: sipper@cs.bgu.ac.il

© Springer International Publishing Switzerland 2016
A. Adamatzky and G.J. Martínez (eds.), *Art of Cellular Automata*,
Emergence, Complexity and Computation 20,
DOI: 10.1007/978-3-319-27270-2_10

2-bit counter ©2015 Moshe Sipper.
The one-dimensional synchronisation task: A 2-bit counter. Operation of a non-uniform, 2-state CA, with connectivity radius $r = 2$. Grid size is $N = 298$. The CA converges to a period-4 cycle upon presentation of a random initial configuration, a behaviour that may be considered a 2-bit counter. [147].

3-bit counter ©2015 Moshe Sipper.
The one-dimensional synchronisation task: A 3-bit counter. Operation of a non-uniform, 2-state CA, with connectivity radius $r = 3$. Grid size is $N = 447$. The CA converges to a period-8 cycle upon presentation of a random initial configuration, a behaviour that may be considered a 3-bit counter [147].

Biological Lattice-Gas Cellular Automata

Andreas Deutsch

Biological lattice-gas cellular automata (LGCA) can be viewed as models for collective behaviour emerging from microscopic migration and interaction processes of biological cells. Such LGCA are used to model the interplay of cells with each other and with their heterogeneous environment by describing interactions at a cell-based (microscopic) scale and facilitating both efficient simulation and theoretical analysis of emergent, tissue-scale (macroscopic) parameters [35]. Historically, LGCA have been introduced as models of gas and fluid flows, through implementing simplistic local collisions. Often, the overall macroscopic behaviour of the system can be approximated very well if averages over larger spatial scales are considered [57]. In a biological context, LGCA particles are interpreted as cells and cell migration is modelled by updating cell positions at each time step based on local cell interactions. Local cell interactions are described by problem-specific LGCA transition rules. These transition rules are different from the rules that have been used for modelling fluid flows. LGCA transition rules in models of cell migration, in general, do not assume energy or momentum conservation. Biological LGCA models can be classified as stochastic cellular automata with time-discrete, synchronous updates consisting of stochastic interaction and subsequent deterministic movement steps. The deterministic movement steps are alternated with stochastic interaction steps, in which processes affecting cell number, e.g., birth and death can be implemented. Implementing movement of individuals in traditional synchronous-update cellular automaton models is not straightforward, as one site in a lattice can typically only contain one individual, and consequently movement of individuals can cause collisions when two individuals want to move to the same empty site. In a lattice-gas model this problem is avoided by having separate channels for each direction of movement and imposing an exclusion principle. In addition, rest channels can be added for non-moving cells.

A. Deutsch
Center for Information Services and High Performance Computing,
Technische Universität Dresden, Germany
e-mail: andreas.deutsch@tu-dresden.de

© Springer International Publishing Switzerland 2016
A. Adamatzky and G.J. Martínez (eds.), *Art of Cellular Automata*,
Emergence, Complexity and Computation 20,
DOI: 10.1007/978-3-319-27270-2_11

The LGCA idea has led to models for migration of individual cells during spatio-temporal pattern formation in microorganisms, cell cultures and developing organisms. The essential modelling idea is the definition of appropriate transition probabilities characterizing specific cell interactions. In particular, cell motion may be influenced by the interaction of cells with components of their immediate local surrounding through haptotaxis or differential adhesion, interaction with the extracellular matrix, contact guidance, contact inhibition, and processes that involve cellular responses to signals that are propagated over larger distances (e.g. chemotaxis). LGCA models have also been used to study emergent collective behaviour in cell swarming [27], angiogenesis [120] and Turing pattern formation. Furthermore, LGCA models have been suggested for various aspects of tumour growth [124]. In particular, simulations and analysis of appropriate LGCA models permit to characterize different growth and invasion scenarios [74, 73].

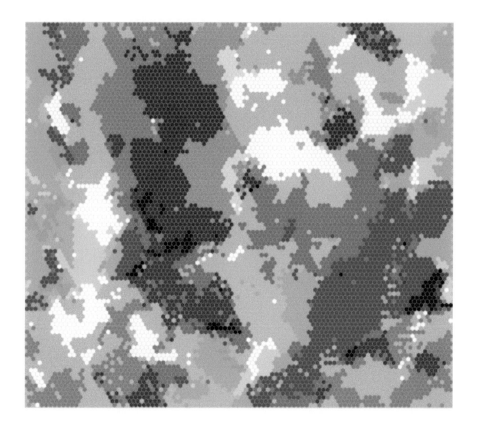

Clusters of alignment in a LGCA simulation. ©2015 Andreas Deutsch.
The figure is based on a LGCA that we have introduced as a model for random walkers with biologically motivated interactions favouring local alignment and leading to collective motion or swarming behaviour [27]. The degree of alignment is controlled by a sensitivity parameter, and a dynamical phase transition exhibiting spontaneous breaking of rotational symmetry occurs at a critical parameter value. The model has been analysed using non-equilibrium mean-field theory. Mean-field predictions have been derived that describe the phase transition as a function of sensitivity and density. Different colours encode different cell orientations. The figure indicates formation of alignment clusters.

The Enlightened Game of Life

Claudio Conti

The interaction of light with complex matter is one of the fundamental subjects in modern physics and biology. This topic is relevant from a variety of different perspectives, including, among others, modelling of animal behaviour (as the moonlight driven coral reef spawning), evolution (the development of the eye), and light activated matter (as, e.g., opto-genetics and laser-driven micro-motors). From a very abstract point of view, if one accepts cellular automata (CA) as the simplest mathematical model for life (as originally suggested by Conway), one can try to include the interaction with light and electromagnetic radiation by enlarging the set of evolution rules to obtain a photo-sensible CA. We follow this approach and add a rule to the Conway's Game of Life CA [30]. In simple words, this new rule states : if you are able to extract energy from light, you survive, independently of the status of the other cells. We also need to include in our model the dynamics of light. Hence, we simulate the evolution of the CA placed inside a cavity filled by electromagnetic radiation. The latter is modelled by the Maxwell equations. The resulting Maxwell/Conway CA show a surprising phenomenology that, so far, has been only marginally investigated. The CA collective motion is able to follow the electromagnetic waves, and display ordered and disordered configurations depending on the strength of the interaction with light. This approach may be also enriched to study if photo-sensible CA may be genetically selected in the presence of competitive non-photo-sensible specimen, or if the interaction of light furnishes novel capabilities to the CA, as, e.g., surviving to catastrophic events and developing large scale organisation. The proposed model may be considered as the simplest approach for light-driven active-matter.

C. Conti
Institute for Complex Systems, National Research Council (ISC-CNR) and Department of Physics, University Sapienza, Rome, Italy
e-mail: claudio.conti@uniroma1.it

© Springer International Publishing Switzerland 2016
A. Adamatzky and G.J. Martínez (eds.), *Art of Cellular Automata*,
Emergence, Complexity and Computation 20,
DOI: 10.1007/978-3-319-27270-2_12

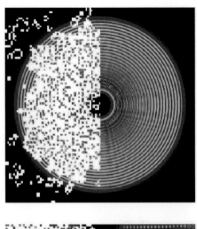

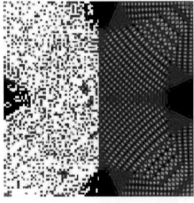

Light driven evolution of CA. ©2015 Claudio Conti.
CA with Conway's Game of Life rules and coupled with a electromagnetic field evolves in a two dimensional cavity. The field is displayed in false colours, the CA is shown in white, only the left-half of the CA is shown. This is an example of an enlightened Game of Life [30].

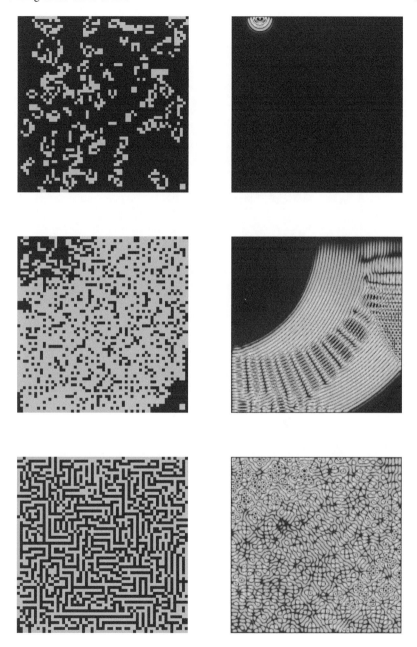

Time evolution ©2015 Claudio Conti.
CA with Conway's Game of Life rules and coupled with an electromagnetic field evolves in a two
dimensional cavity. Different snapshots are shown with a wave generated in the cavity and the CA
evolving accordingly, times increases from top to bottom. Left panels show the CA configurations,
right panels show the electromagnetic wave. [30].

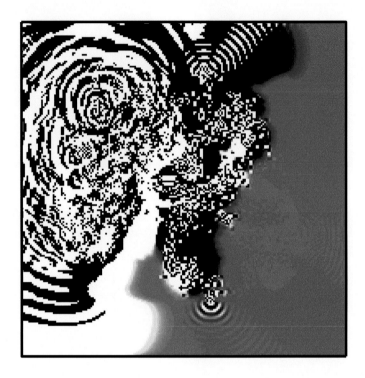

Nonlinear evolution ©2015 Claudio Conti.
Snapshot of the electromagnetic wave scattered by a CA that absorbs light. The nonlinear feedback action of the CA on the electromagnetic wave is included in the simulation. [30].

Small Synchronizers and Prime Generators

Hiroshi Umeo

We illustrate a prime sequence generation problem and a firing squad synchronisation problem on cellular automata (CA) with different communication models. For a long time there was little use of prime numbers in practical applications. But nowadays, it has been known that large scale prime numbers play a very important role in encryption in computer security networks. A question is "How can we generate prime numbers in real-time on a small-state CA?". We present two implementations of real-time prime generators on the CA having smallest, known at present, number of internal states. It is shown that there exists a real-time prime generator on 1-bit inter-cell communication CA with 25-states, which is an improvement over a 34-state implementation given in [152]. A infinite prime sequence can be also generated in real-time by an eight-state CA with constant-bit local communications. The algorithms illustrated are based on the classical sieve of Eratosthenes, and our eight-state implementation is an improvement over a nine-state prime generator presented in [81]. Those two implementations on CA with different communication models are the smallest realisations in the number of states, at present, see [156].

The synchronisation in ultra-fine-grained parallel computational model of CA has been known as the firing squad synchronisation problem (FSSP) since its development, in which it was originally proposed by Myhill in the book edited by Moore [123] to synchronise all or some parts of self-reproducing cellular automata. A rich variety of synchronisation algorithms has been proposed [150]. An informal definition of the FSSP is as follows. Consider a finite one-dimensional (1D) cellular array consisting of n cells. Each cell is an identical (except the border cells) finite-state automaton. The array operates in lock-step mode in such a way that the next state of each cell (except border cells) is determined by both its own present state and the present states of its left and right neighbours. All cells (*soldiers*), except the left end cell (*general*), are initially in the quiescent state at time $t = 0$ with the property that the next state of a quiescent cell with quiescent neighbours is the

H. Umeo
Univ. of Osaka Electro-Communication, Nayagawa-shi, Hastu-cho, 18-8, Osaka, 572-8530, Japan
e-mail: umeo@cyt.osakac.ac.jp

© Springer International Publishing Switzerland 2016
A. Adamatzky and G.J. Martínez (eds.), *Art of Cellular Automata*,
Emergence, Complexity and Computation 20,
DOI: 10.1007/978-3-319-27270-2_13

quiescent state again. At time $t = 0$, the left end cell C_1 is in the *fire-when-ready* state, which is the initiation signal for the array. The firing squad synchronisation problem is to determine a description (state set and next-state function) for cells that ensures all cells enter the *fire* state at exactly the same time and for the first time. The set of states and the next-state function must be independent of n. Here we present some variants of the FSSP solutions with smallest state implementations, known at present: an eight-state optimum-time generalised synchroniser, a 35-state optimum-time synchroniser for 1-bit inter-cell-communication CA, a seven-state square synchroniser, a 12-state rectangle synchroniser, and a minimum-four-state, minimum-time partial synchroniser for rings.

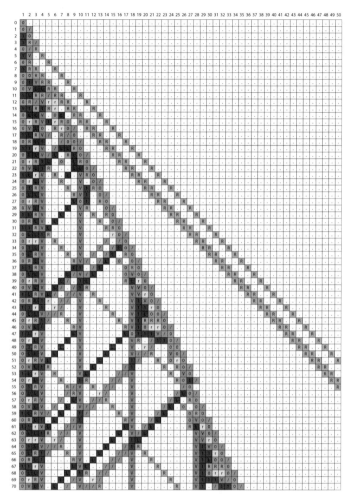

An eight-state real-time prime generator on conventional $O(1)$-bit communication CA.
© 2015 Hiroshi Umeo, Kunio Miyamoto, and Yasuyuki Abe.

We have a semi-infinite 1D CA M consisting of identical finite state automata. Let $\{t_n \mid n = 1, 2, 3, ...\}$ be an infinite monotonically increasing positive integer sequence defined on natural numbers such that $t_n \geq n$ for any $n \geq 1$. We say that M can generate a sequence of $\{t_n \mid n = 1, 2, 3, ...\}$ in real-time if and only if the leftmost end cell of M falls into a special state at time $t = t_n$. Here we explore a real-time prime sequence generation problem and present two implementations of real-time prime generators on CA having smallest number of internal states, known at present. We show that an infinite prime sequence can be generated in real-time by an eight-state CA with $O(1)$-bit local communications. The algorithm used is based on the classical sieve of Eratosthenes, and our eight-state implementation is an improvement over a nine-state prime generator presented in [81], see [156] for details.

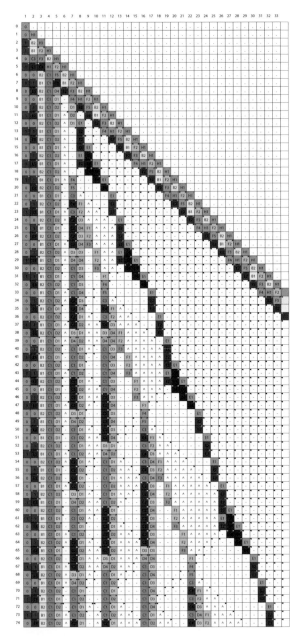

A 25-state real-time prime generator on 1-bit inter-cell communication CA. ©2015 Hiroshi Umeo, Kunio Miyamoto, and Yasuyuki Abe.

It is shown that there exists a real-time prime generator on 1-bit inter-cell communication CA with 25-states. The result is an improvement over a 34-state implementation given in [152]. These two implementations on CA with different communication models are the smallest realisations in the number of states. See [156].

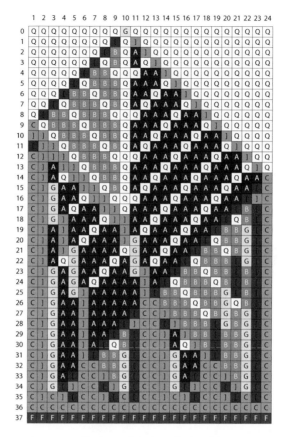

An 8-state optimum-time generalized synchronizer ©2015 Hiroshi Umeo, Naoki Kamikawa, Kouji Nishioka, and Shunsuke Akiguchi

The synchronisation in CA has been known as firing squad synchronisation problem (FSSP) since its development, in which it was originally proposed by Myhill in [123] to synchronise all parts of self-reproducing CA. The problem has been studied extensively for more than fifty years. The original FSSP is stated as follows. Given an array of n identical CA, including a general on the left end which is activated at time $t = 0$, we want to give the description (state set and next-state transition function) of the automata so that, at some future time, all of the cells will simultaneously and, for the first time, enter a special firing state. The initial general is on the left end of the array in the original FSSP. The generalized FSSP is an extended version which allows the initial general to be located at any cell. The tricky part of the problem is that the same kind of soldier having a fixed number of states must be synchronised, regardless of the position of the general and the length n of the array. The eight-state optimum-time generalised implementation [153] is the smallest one, known at present.

A 35-state optimum-time synchroniser for 1-bit inter-cell communication CA ©2015 Hiroshi Umeo and Takashi Yanagihara.

Here we study the FSSP on 1-bit inter-cell communication CA. The 1-bit communication CA is a weakest subclass of CA in which the amount of inter-cell communication bits transferred among neighbouring cells at one step is restricted to 1-bit. We show a 35-state implementation of optimum-time FSSP algorithm. The implementation proposed is the smallest one, known at present. See [157].

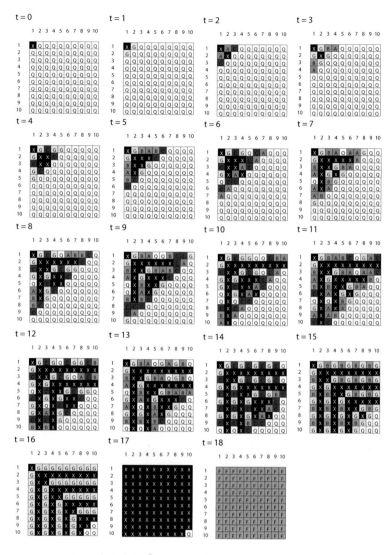

A seven-state square synchronizer ©2015 Hiroshi Umeo and Keisuke Kubo.
Here we show a seven-state optimum-time synchronisation algorithm that can synchronise any square arrays of size $n \times n$ with a general at one corner in $2n - 2$ steps, which is a smallest realisation of time-optimum square synchroniser. The implementation is based on a zebra-like mapping schema which embeds synchronisation operations on 1D arrays into square arrays. See [155].

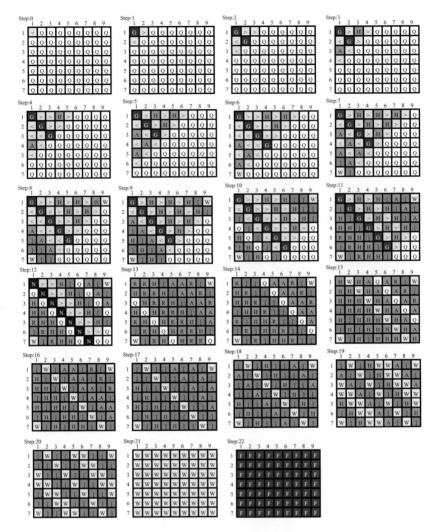

A 12-state rectangle synchroniser. ©2015 Hiroshi Umeo, Masaya Hisaoka, and Shunsuke Akiguchi.

Here we show a new optimum-time synchronisation algorithm that can synchronise any 2D $m \times n$ rectangular arrays in $m + n + max(m,n) - 3$ steps. The algorithm is based on a state-efficient diagonal mapping schema for embedding a special class of generalised 1D optimum-time synchronisation algorithms onto 2D rectangular arrays. The implementation is given on a 12-state CA. See [151].

A four-state partial solution to ring FSSP ©2015 Hiroshi Umeo, Naoki Kamikawa, and Jean-Baptiste Yunẽs.

We have given a family of smallest symmetric four-state firing squad synchronisation protocols that can synchronise any 1D ring cellular arrays of length $n = 2^k$ for any positive integer k. Note that the number four is the smallest one in the class of partial synchronisation protocols proposed so far. The solution can synchronise infinite number of cells, but not all. Several solutions are based on Wolfram's rules 60 and 150.

Ecological Patterns of Self-Replicators

Hiroki Sayama

The aesthetic sense of humans has part of its origins in human evolution. Our ancestors must have evolved the abilities to feel various aesthetic values on other living creatures they shared their habitats with, sometimes with joy, sometimes with fear, sometimes with awe. The same principle might apply to the art of cellular automata (CA) too; if you see living creatures flourishing or struggling in CA, that might trigger a certain kind of aesthetic feeling that is rather different from the one you feel for purely mathematical, geometrical patterns.

Artificial Life is a research field where CA are often used to create ecological and evolutionary dynamics of virtual self-replicating creatures. In this short section, some of my work in this field are displayed. They are all based on purely mathematical dynamics developing on a regular spatial grid, yet one can see ecologies and evolution of self-replicating life form as an emergent property of those systems. Such biological looking behaviours of CA can surely trigger certain kinds of feelings to those who watch them. One of the YouTube users who watched a movie of evoloops (creatures shown in the first two figures) once left the following critical comment[1]:

> *"I didn't even pay attention and I can tell the little cells were screaming "help, we die and we don't know what else to do", how can you stand there and say nothing while this torture is going on?"*
> *– MrGodKid (2012)*

This kind of reaction was precisely what I wanted to induce to people by creating various life forms in CA and other computational media. It is part of the constructive approaches taken by the Artificial Life researchers, seeking potential answers to the fundamental question: *What is life?*

You can find more details of each piece of work in the references cited in the respective caption.

H. Sayama
Binghamton University, P.O. Box 6000, Binghamton, NY 13902-6000, USA
e-mail: sayama@binghamton.edu

[1] www.youtube.com/watch?v=vbpoTZlNTiw

© Springer International Publishing Switzerland 2016
A. Adamatzky and G.J. Martínez (eds.), *Art of Cellular Automata*,
Emergence, Complexity and Computation 20,
DOI: 10.1007/978-3-319-27270-2_14

Ecology and evolution of evoloops. ©2015 Hiroki Sayama.
This evoloop CA was derived from Chris Langton's self-reproducing loops [87]. Each cell of the evoloop automaton takes nine states with von Neumann neighborhoods. Genetic information flows counter-clockwise inside the sheath structures made of state-2 (red) cells, which will be sent outward through extended "arms" and then translated into new sheath structures at their end points. Physical collision between multiple loops causes appearance of irregular spatial patterns, which is the primary source of mutation in this deterministic universe. In the screen shot on the left, internal sheaths are colored according to the "species" (size and genotype) of each loop, whose detailed breakdowns are shown on the right [137, 138]. Simulation software is available online at `www.necsi.edu/postdocs/sayama/sdsr/#software`.

Large-scale simulation of evoloops in a hostile environment. ©2004 Chris Salzberg, Antony Antony, Hiroki Sayama.
This shows a snapshot of a large-scale simulation of the evoloop CA on a 3,000 × 3,000 grid with infectious "disease" states involved (shown in gray) [136]. Uninfected evoloops are coloured differently based on their species (size and genotype). Movies are available online at http://artis.phenome.org/.

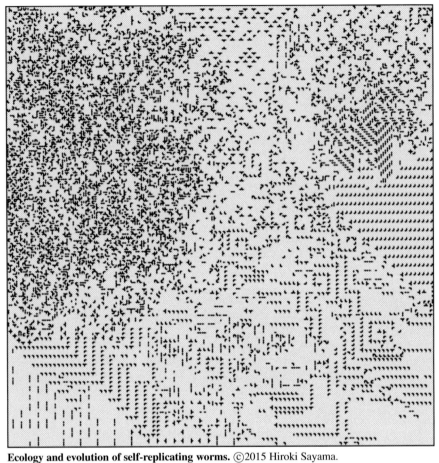

Ecology and evolution of self-replicating worms. ©2015 Hiroki Sayama.
Each cell of this CA contains genetic information about branching structure of organisms, which will be transmitted across their boundaries when they collide [139]. The image was post-processed with an image filter to add aesthetic effects. Simulation software is available online at www.necsi.edu/postdocs/sayama/worms/#applet.

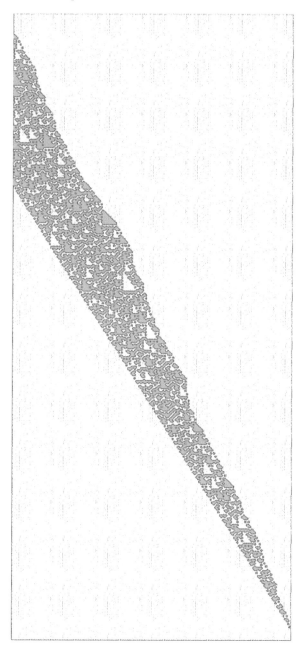

Trajectory of the Collatz sequence. ©2015 Hiroki Sayama.
This is not quite CA, but instead shows the trajectory of the Collatz sequence from an initial value
12345678910111213141516171819120, visualised using the method described in [140]. While the
Collatz sequence is defined purely mathematically, one can still interpret its behaviour as an outcome of the competition between two distinct ecological processes — growth and extinction of
self-replicating cells [140]. The image was post-processed with an image filter to add aesthetic
effects.

The Art of Penrose Life

Susan Stepney

John Horton Conway's Game of Life (GoL) [26, 106] is a simple two-dimensional, two state cellular automaton (CA), remarkable for its complex behaviour [26, 133].

The classic GoL is defined on a regular square lattice. The update rule depends on the state of each cell and its neighbouring eight cells with which it shares a vertex. Each cell has two states, 'dead' and 'alive'. If a cell is alive at time t, then it stays alive if and only if it has two or three live neighbours (otherwise it dies of 'loneliness' or 'overcrowding'). If a cell is dead at time t, then it becomes alive (is 'born') if and only if it has exactly three live neighbours. This rule gives a famous zoo of GoL patterns, including still lifes, oscillators, and gliders.

Here we show some results of running GoL rules on Penrose tilings. More detail can be found in [127], from which all the figures here are taken. The neighbourhood of a Penrose tile is again all the tiles with which it shares a vertex; now there can be 7–11 of these, depending on details of the tiling. We show some interesting still life patterns and oscillator patterns. For a fuller, but still preliminary, catalogue of Penrose life structures, see [127]. These patterns were discovered by a combination of systematic construction and random search.

S. Stepney
University of York, UK, YO10 5DD
e-mail: susan.stepney@york.ac.uk

© Springer International Publishing Switzerland 2016
A. Adamatzky and G.J. Martínez (eds.), *Art of Cellular Automata*,
Emergence, Complexity and Computation 20,
DOI: 10.1007/978-3-319-27270-2_15

Fig. 1 A kite and dart tiling shaded by the eight distinct neighbourhood types; neighbourhood sizes range from 8–10; darker tiles have more neighbours. Reproduced from [127, fig.18.11].

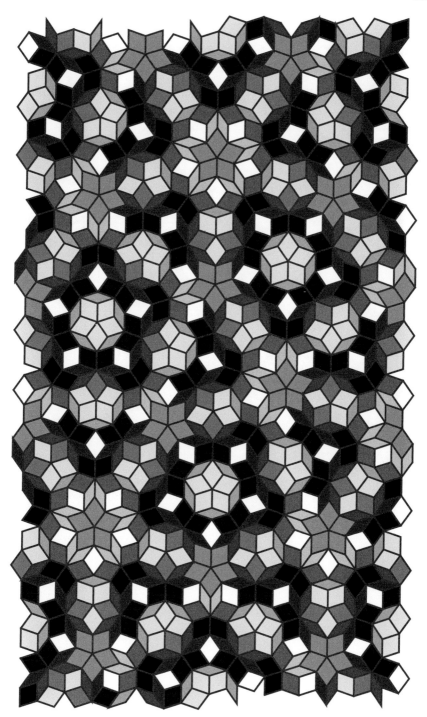

Fig. 2 A rhomb tiling shaded by the 11 distinct neighbourhood types; neighbourhood sizes range from 7–11; darker tiles have more neighbours. Reproduced from [127, fig.18.13].

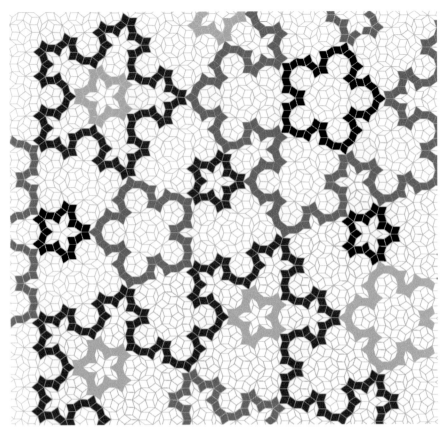

Fig. 3 Arbitrarily large still lifes can be constructed on the rhomb tiling. Pick a thick rhomb (of a particular type; see [127] for details). Complete the "ribbon" of thick rhombs that is formed from the two thick rhombs adjacent to its edges (all thick rhombs have precisely two such thick rhomb neighbours). The figure shows several such ribbons of thick rhombs; different grey shades show different possible ribbons, although these cannot all exist simultaneously. Reproduced from [127, fig.18.38].

Fig. 4 The "bat-to-bat" oscillator. Live cells at each step are shown in black; cells that are dead this step but live some other step are shown in grey; cells that are always dead are shown in white. This is an oscillator on the kite-and-dart tiling; it has period 4; it has a total of 28 cells live at some step in the oscillator; it has a minimum of 12 live cells during its oscillation (at steps 1 and 3). It is given the code kd-p4-28-12. Reproduced from [127, fig.18.71d].

Fig. 5 The "moustaches" oscillator. This is an oscillator on the kite-and-dart tiling; it has period 9; it has a total of 36 cells live at some step in the oscillator; it has a minimum of 10 live cells during its oscillation (at step 4). It is given the code kd-p9-36-10. Reproduced from [127, fig.18.82].

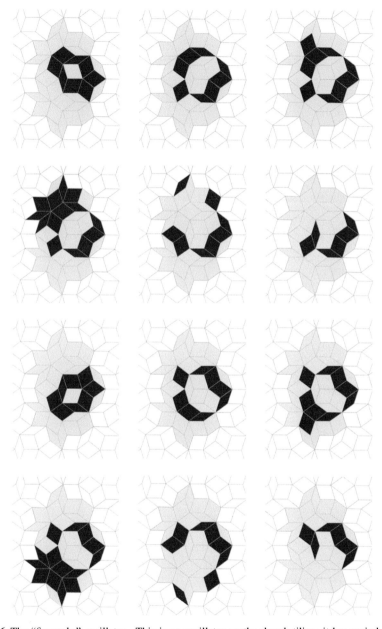

Fig. 6 The "fireworks" oscillator. This is an oscillator on the rhomb tiling; it has period 12; it has a total of 33 cells live at some step in the oscillator; it has a minimum of 6 live cells during its oscillation (at steps 6 and 12). It is given the code r-p12-33-6. Note the underlying period 6 behaviour, combined with a reflection. Reproduced from [127, fig.18.96].

Asynchronous Cellular Automata Simulating Complex Phenomena

Olga Bandman

The concept of asynchronous cellular automaton (ACA) differs from that of classical synchronous cellular automata (CA) in the mode of operation, namely cells tran-
sit to next states at random being updated sequentially. ACA are especially suitable
for simulation chemical, biological, and physical phenomena, which are dissipative,
nonlinear, and stochastic. Usually, such phenomena are represented as a combination of movements and transformations of real or abstract particles. In simulation
practice such processes are combined under a common concept of reaction-diffusion
processes. Using the formalism of [1] we define ACA by three notions: (i) a state
alphabet A, (ii) a finite set X of cell coordinates in a discrete space, and (iii) a global
operator $\Theta(X)$. The latter transfers the current global configuration of the ACA $\Omega(t)$
into the next state $\Omega(t+1)$, t being considered as iteration number in the ACA evolution. In general, global operator is a composition [22] of several probabilistic local
operators, sometimes referred to as transition rules, i.e. $\Theta(X) = \Phi(\theta_1(x), \dots \theta_n(x))$,
resulting in application of all $\theta(x)$ to all $x \in X$ in the order being prescribed by Φ.
Local operators are defined by a cell neighborhood $T(x) = \{x, x_1, \dots, x_m\}$, and a
set of transition functions $u'_j = f_j(u, u_1, \dots, u_m)$, u_j being the current state of a cell
$x_j \in T(x)$. Application of $\theta(x)$ to a cell $x_k \in X$ consists of computing next states u'_j,
of all cells in the neighborhood of x_k according to their transition functions, and immediately replacing current states by the new ones. When the cell states are symbols
or constants, what is typical for ACA models of chemical and biological processes,
transition functions degenerate into states substitutions or interchanging, $\theta(x)$ being
then referred to as a substitution.

At contrast to synchronous CA, a transition rule in ACA is allowed to update several cell states simultaneously. It does not cause any collision when being computed
on a sequential computer because of the sequential choice of cells to be changed.
However, dangerous collisions appear when a large scale ACA is allocated on sev-

O. Bandman
Institute of Computational Mathematics and Mathematical Geophysics, SBRAS pr. Lavrentieva,
6, Novosibirsk, 630090, Russia
e-mail: bandman@ssd.sscc.ru

© Springer International Publishing Switzerland 2016
A. Adamatzky and G.J. Martínez (eds.), *Art of Cellular Automata,*
Emergence, Complexity and Computation 20,
DOI: 10.1007/978-3-319-27270-2_16

eral processors of a parallel supercomputer. In that case in order to avoid collisions, each updating of a cell, whose neighbors occur in the adjacent processor should be immediately transmitted to it, which might have lead to extremely low parallelization efficiency. The problem was solved by performing a transformation of ACA into an equivalent block-synchronous CA. The transformation induces a certain amount of synchronism reducing the amount of data interprocessor transmissions to $m = |T|$, which is quite acceptable [21].

The real life complex phenomena are usually simulated by composed ACA [22], the following types of composition being mostly used:

(i) local superposition, where for each randomly chosen $x \in X$ a superposition of all θ_i from $\Theta(X)$ is applied,

(ii) global superposition, or superposition of several $\Theta_i(X)$, which themselves may be simple or composed,

(iii) parallel composition, called also ACA systems, is used when $n > 1$ species are involved in the process under simulation. The composition may be regarded as an ACA system consisting of n component ACA, or as a single composed ACA with a global operator $\Theta(X) = \Psi(\Theta(X^{(1)}),\dots,\Theta(X^{(n)}))$. In both cases the cellular space X is represented in the form of several layers, $X = X^{(1)} \cup \dots \cup X^{(n)}$, a cell $x_i \in X$ being a set of corresponding cells in all layers, i.e. $x_i = \{x_i^{(1)},\dots,x_i^{(n)}\}$. Each $\Theta_i(X)^{(i)}$ updates the cells from its own $(X)^{(i)}$, the transition rules being defined on the whole array [24].

Three typical examples of composed reaction-diffusion ACA are presented below. The first example shows a parallel composition of two simple operators. One of them (simulating redistribution of heat) operates independently, while the other (simulating pattern formation) uses the neighborhood consisting of cells in both layers. The example illustrates the simulation of a process, whose properties are dynamically changed both in time and in space by another process [23]. The second example shows a parallel composition of two interdependent global operators, simulating interactions between prey and predator, each being a local superposition of diffusion (simulating movements) and reaction (simulating species population) [24]. Simulation results showed, that the system exhibits self organization which results in forming stable dense spots of both species. The third example is an illustration of a complex large–scale 3D ACA, global operator being an hierarchical composition, a higher level of which is a global superposition of six global operators, which, in their turn, are local superpositions of substitutions. The size of the ACA is $1.4 \cdot 10^9$ cells, which required the simulation to be performed on a 49 processor of a multicomputer cluster [25].

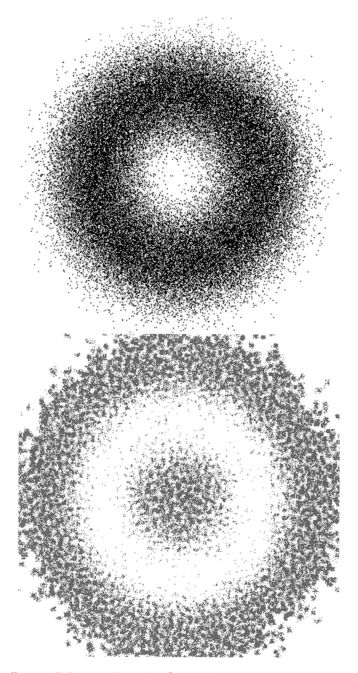

Dynamically controlled pattern formation. ©2015 Olga Bandman.
Two coupled asynchronous cellular automata (CA) — asynchronous totalistic automaton and diffusion automaton — operate on a shared non-uniformly heated chemically active substrate. The inhibiting interaction between the automata is clearly visible in the pictures [22].

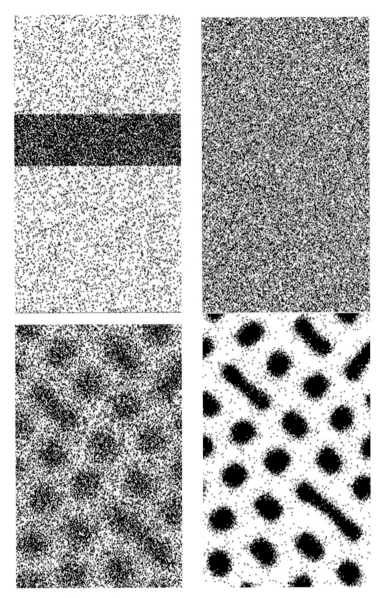

Preys and predators. ©2015 Olga Bandman
These are configuration of two cellular automata coupled in a prey-predator fashion. These asynchronous probabilistic automata imitate diffusion of two species and 'reaction' between the species. The reaction rules simulate predator and prey interactions. The evolution is a self-organizing process that terminates in the stable state after hundred of iterations. Difference between stable states patterns is caused by the difference in diffusion coefficients: predator in more agile, while prey is concentrated in compact spots. Simulation was implemented on two cores of a multi core computer with a shared memory [24]. Pattern of predators is on the left, pattern of preys is on the right.

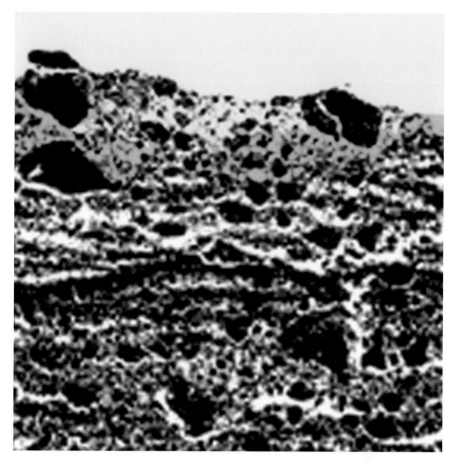

Water permeation in soil. ©2015 Olga Bandman.
The CA model for simulating water permeation through porous soil with complex morphology. Cell states signify air, water, hard substance an hydrophilic inclusions. The automaton transitions rules simulate water particles propagation, soaking of hydrophilic pores, drainage from soaked cell during the evaporation. The 3D automaton of $700 \times 700 \times 1480$ cells was implemented on 49 processors of a supercomputer. To obtain acceptable efficiency of parallelization the asynchronous operators were transformed into block-synchronous mode [25].

A Multiparticle Lattice-Gas Cellular Automaton Simulating a Piston Motion

Yuri Medvedev

The well known lattice-gas cellular automata (CA), [38] simulate incompressible viscous fluid flow. The cell states contains Boolean six component vectors $s(x) = (s_1, \ldots, s_6)$, where $s_i = 1$ is interpreted as a particle with a unit mass, propagating towards the ith cell of the neighborhood $T(x)$, which contains six adjacent to x cells in a *hexagonal cellular array*. Global transition rule $\Theta(X)$ is a superposition of two global operators: $\Theta_C(X)$, simulating particles collision, and $\Theta_P(X)$, simulating the propagation. Transition rules obey mass and momentum conservation laws. Simulation capabilities of the model are limited by the maximal Reynolds number, which does not allow to use the model for simulating inviscid gas flows. Moreover, in the frames of the model it is not possible to simulate the flow passing around a moving obstacle. These limitations stimulated to creation of a modification, called multi-particle lattice gas CA [118], where Boolean state vectors are replaced by similar vectors with integer components, thus allowing a finite number of particles to propagate towards a neighboring cell. A main characteristic of the model is a maximal cell population $N_{max} = \sum_i^6 s_i'$, where s_i' are integers. The iteration consists of two phases: at the first phase the propagation local transition rules are applied to all cells, displacing $s_i'(x)$ particles from all cells $x \in X$ to their ith neighbors. The collision rules redistribute particles between the cell state vector components, preserving the condition of conservation laws.

Such expansion of the model essentially, widened its applicability domain increasing maximal Reynolds number and allowing to simulate gas behavior with non-stationary boundary conditions. These modifications required to develop a new set of transition rules, whose quantity depends on N_{max}, as well as to develop the method of constructing transition rules in the vicinity of moving obstacles [119]. The price for these capabilities is the significant increase of computational time for collision rules implementation. A bright practical example of gas flow with non-stationary boundary conditions is a work of a piston going back and forth in a pipe. Simulation results of the process showed the qualitative correspondence with the laws of Toricelli, Pascal and Poiseuille [119].

Y. Medvedev
Institute of Computational Mathematics and Mathematical Geophysics, SBRAS pr. Lavrentieva, 6, Novosibirsk, 630090, Russia
e-mail: medvedev@ssd.sscc.ru

© Springer International Publishing Switzerland 2016
A. Adamatzky and G.J. Martínez (eds.), *Art of Cellular Automata*,
Emergence, Complexity and Computation 20,
DOI: 10.1007/978-3-319-27270-2_17

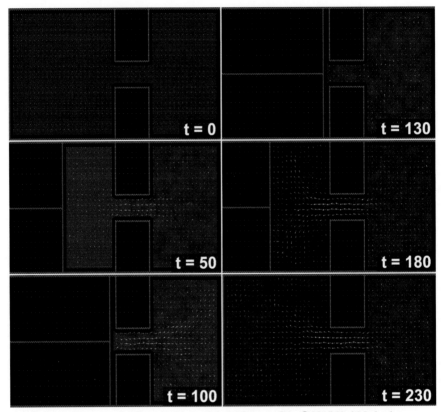

Simulating a piston motion by a multi-particle lattice-gas CA. ©2015 Yuri Medvedev.
A piston motion is simulated by a multi-particle lattica-gas CA [118] The piston moves in a reservoir consisting of two chambers, connected by a diffuser and filled by gas. The cell states set contains integer vectors of length $n = 6$. It operates synchronously on a cellular array of size 100×200 hexagonal cells. Maximal cell population $N_{max} = 20$. Some snapshots of computer simulation of piston moving back and forth in the left chamber of the pipe are shown in the picture. High pressure is visualized in bright red, low pressure in dark red, vacuum is black. At $t = 0$ the piston starts to move rightwards towards the diffuser, gas being gradually compressed ($t = 50$) between the diffuser and the piston. At $t = 100$ the piston stops letting the gas precipitate though the diffuser aperture. At $t = 130$ the piston starts moving back ($t = 180$) under the pressure difference, until at $t = 230$ the system comes into the initial state [119].

Two Layer Asynchronous Cellular Automata

Anastasiya Kireeva

Asynchronous cellular automata (ACA) whose transition functions contain activation and inhibition components are capable to simulate a great variety of complex systems behavior. The competition between activators and inhibitors impact on the ACA functioning contributes to regulation of ACA evolution properties [174]. Particularly, activator/inhibitor ratio determines the type of ACA stable behavior: a stable state in the form of a constant pattern, stable oscillations, propagating autowaves, or chaos. Moreover, by changing activator/inhibitor parameters in time and/or in space, the process under simulation may be tuned to capture certain properties of the phenomenon under simulation. Such capabilities are especially helpful when ACA simulation aims to study chemical or biological processes on micro or molecular level. When the process under simulation proceeds in changing conditions, a composition of two ACA [22] should be used, referred to as a two layer ACA. The additional layer being an ACA, whose evolution simulates the variation of the main ACA parameters. Transition functions of the main ACA use the cells states in the second layer as variables. In chemistry the role of activators and inhibitors is usually played by positive and negative reaction rates, respectively, which is shown in the investigation of carbon monoxide (CO) oxidation reaction on catalyst in the gas mixture $CO + O_2$ using Monte-Carlo method [62]. The similar reaction has been also studied using ACA [40]. The need to find out how the reaction proceeds in changing temperature conditions motivated to use an additional ACA in the form of the second layer [80].

A. Kireeva
Institute of Computational Mathematics and Mathematical Geophysics, SBRAS pr. Lavrentieva, 6, Novosibirsk, 630090, Russia
e-mail: kireeva@ssd.sscc.ru

© Springer International Publishing Switzerland 2016
A. Adamatzky and G.J. Martínez (eds.), *Art of Cellular Automata*,
Emergence, Complexity and Computation 20,
DOI: 10.1007/978-3-319-27270-2_18

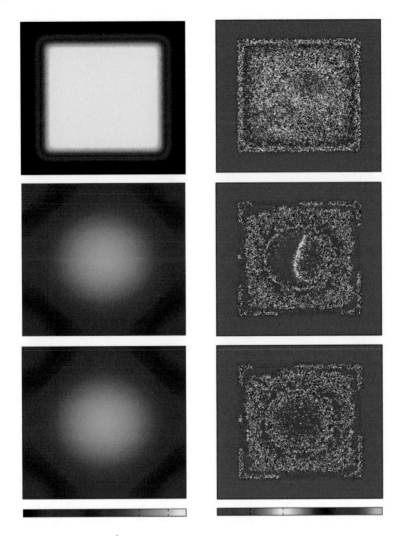

Catalytic reaction at dynamically varying temperature. ©2015 Anastasiya Kireeva.
A two-layer ACA simulates a carbon monoxide (*CO*) oxidation reaction allowing for non-uniform surface temperature distribution. During the reaction different surface waves and patterns appear being invoked by periodical reconstruction of crystal surface from a hexagonal to a cubic structure and back under the influence of the adsorbed *CO*. Surfaces with hexagonal and cubic crystal structures possess different catalytic activity, having the effect of activation and inhibition. Therefore, the reconstruction of surface structure causes oscillations of reagents concentrations. Moreover, cooling and heating of some parts of the surface impact on the reaction dynamics [88, 117]. The figure shows the reagent distributions over the catalyst surface with initially heated central square (480 Kelvin) and cold boundaries (200 Kelvin). At the beginning, alternating *CO* and O_2 covering of the heated central square occurs. With cooling down of initially heated square due to the heat transfer, oxygen ring appears on the boundaries of this square [80]. Left plates are temperature, right plates are distribution of reagents. Red and pink colors mark *CO*, yellow and brown colors mark empty surface with cubic and hexagonal crystal structures, blue color marks oxygen.

Cellular Automata Simulation of Bacterial Cell Growth and Division

Anton Vitvitsky

The exact mechanisms of growth and division in bacterial cells (further, bacterial cell is referred to as bacterium to avoid confusion with a cell in cellular automata) is not clear so far. Investigations in the field are focused on studying the process in the bacteria of *Escherichia coli* (E.coli) [48, 89]. Currently, it is stated that growth and division of a bacterium are invoked by self-organization in *MinCDE* protein system, whose components (*MinC*, *MinE*, *MinD* proteins) interact on the inner surface of the bacterium membrane. The investigations of these processes are of two kinds: experimental study of proteins dynamics *in vitro* [90], and microscopic investigation of the process *in vivo* [91].

In vitro studies of *MinCDE* system revealed the existence of autowaves and some time-spatial patterns, and it was suggested that they arise from an interplay of two opposing mechanisms: cooperative binding of *MinD* proteins to the bacterium membrane, and their accelerated detachment due to the accumulation of *MinE* proteins on the membrane. Based on these assumptions the asynchronous cellular automaton (ACA) model of the *MinDE* behavior *in vitro* is defined, operating on a 2D Cartesian cellular array, using a cell alphabet containing protein names $A = \{MinD, MinE, MinDE\}$, and the transition rule, which is a global superposition [22] of several simple operators mimicking elementary interactions between protein abstract particles and between ones and the membrane.

To simulate the process *in vivo* the set of transition rules in the superposition is composed with the set of rules, simulating growth and division of the bacterium. To construct such complex CA model three following problems were solved: a method for designing mathematical representation of the cellular array X imitating the 3D surface of the bacterium membrane is developed [161]; transition functions on X simulating growth of the array in length, and division it in two parts are determined [160], and a method for modifying transition rules designed for Cartesian cellular array into equivalent ones for operating on X is elaborated.

A. Vitvitsky
Institute of Computational Mathematics and Mathematical Geophysics, SBRAS pr. Lavrentieva, 6, Novosibirsk, 630090, Russia
e-mail: vitvit@ssd.sscc.ru

© Springer International Publishing Switzerland 2016   121
A. Adamatzky and G.J. Martínez (eds.), *Art of Cellular Automata*,
Emergence, Complexity and Computation 20,
DOI: 10.1007/978-3-319-27270-2_19

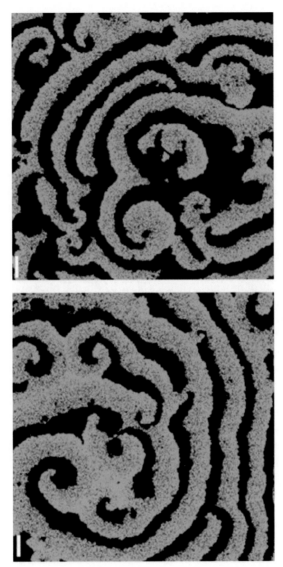

Autowaves in bacterial MinCDE protein system. ©2015 Anton Vitvitsky.
Recent studies of MinCDE system behavior *in vitro* showed the emergency of protein auto-waves
formation on the inner surface of bacteria membrane. The process is simulated by ACA, whose
cell states are names of interacting proteins *MinE*, *MinD*, *MinDE*, and a vacant sites of the membrane. A transition rule is a global superposition of elementary transition rules simulating the
following actions: attachment of *MinD* to the membrane from the cytoplasm; binding of *MinE* to
the membrane-bound *MinD*; diffusion of *MinDE* and *MinD* over the membrane; decomposition
of *MinDE* as a result of hydrolysis; detachment of *MinD* and *MinE* from the membrane;' and
rebinding of membrane *MinE* with membrane-bound *MinD*. All transition rules are probabilistic. Variation of the probabilities allows to obtain autowaves of different patterns. Visualisation of
computational experiments shows emergency of protein waves, that are similar to those obtained
in vitro experiments [162].

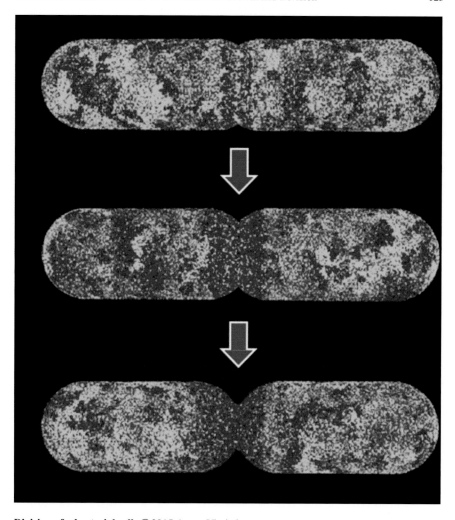

Division of a bacterial cell. ©2015 Anton Vitvitsky.

Growth and division of the bacterium *E. coli* are invoked by complex self-organization mechanism on the inner surface of the membrane. Interaction of *MinD* and *MinE* proteins gives rise
to autowaves propagating from pole to pole of bacterium (in the snapshots *MinD* and *MinE* proteins particles being shown in blue and pink, respectively). For constructing a discrete cellular
array X having the form of the bacterium surface, a special method is developed. The membrane
is represented as a hollow cylinder with semi-spheroids at both ends of X, constructed of rings
$X = R_1 \cup \ldots, R_L$, each R_i consisting of identical cells, centers of all rings being allocated on the
cylinder axis. The operator of growth operator of growth θ_G inserts an additional ring between two
rings in the middle part of bacterium length, which yields in adding a new ring name in the set of
rings. The operator of division θ_D reduces rings radii in some adjacent rings in the central part of
bacteria. Thus, the changing of the ACA cellular array structure dynamically is performed without
reconfiguration of the whole array, but with the necessity to modify the transition rules applied to
the cells, whose neighbourhood is changed due to the reconfiguration of the array [161].

Seismic Cellular Automata

Ioakeim G. Georgoudas, Georgios Ch. Sirakoulis,
Emmanuel M. Scordilis, and Ioannis Andreadis

The proposed cellular automata (CA) driven potential-based model for earthquake simulation is a dynamic system constituted of cells-charges. It is assumed that the system balances through the exercitation of electrostatic Coulomb-forces among charges, without the existence of any other form of interconnection in-between. Such kinds of forces are also responsible for this level to be bonded with a rigid but moving plane below. Provided that at time $t = 0$ the potential $V_{(i,j)}$ of a cell placed at (i, j) exceeds the threshold value V_{th} of the level below, the balance is disturbed and the cell is moved. The moving cell is transferred to a point of lower energy, hence at a state of lower but nonzero potential. The removal of the cell reorders the values of the potential at its nearest neighbours. Potential values' conversion results into new cells to become unstable driving to the appearance of the well known as cascade phenomenon, which is the earthquake's equivalent of the proposed model. The effectiveness and the reliability of the model have been detected by different kinds of measurements like Critical state: the state of the system after a large number of earthquake simulations, Cascade (earthquake) size: the total number of cells that participate in a single earthquake procedure, which stands as long as the condition $V_{(i,j)} > V_{th}$ is true (a measure of the total energy released during the evolution of the earthquake, a measure of the earthquake's magnitude) and Path to criticality: visualisation of the path, which leads from disorder to an order (self-organised criticality). The model has been tested and calibrated with the use of real data.

I.G. Georgoudas · G.Ch. Sirakoulis · I. Andreadis
Laboratory of Electronics, Department of Electrical and Computer Engineering,
Democritus University of Thrace, Greece
e-mail: {igeorg,gsirak,iandread}@ee.duth.gr

E.M. Scordilis
Department of Geophysics, School of Geology, Aristotle University, Greece
e-mail: manolis@geo.auth.gr

© Springer International Publishing Switzerland 2016   125
A. Adamatzky and G.J. Martínez (eds.), *Art of Cellular Automata*,
Emergence, Complexity and Computation 20,
DOI: 10.1007/978-3-319-27270-2_20

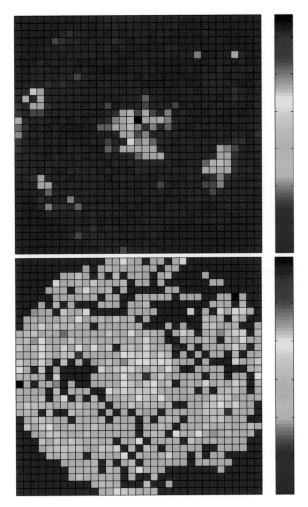

Modeling earthquake activity with CA. ©2011 I.G. Georgoudas, G.Ch. Sirakoulis, E.M. Scordilis and I. Andreadis. Real data is introduced to the model since each examined area has been divided into cells (each cell has been set to 5 km, and the accuracy in the estimation of the focal parameters of the earthquakes (epicentre, focal depth) is of the order of 5 km). Initially, earthquake events are allocated to certain cells of the CA lattice, based on their geographical co-ordinates, following a simplification rule, according to which a 5 km step roughly corresponds to a 0.05 degree step both in the latitudinal and in the longitudinal direction. At the next stage, for each cell, the number of events included in its area is computed and the maximum magnitude is detected. Characteristic example of such distributions is depicted in the top plate where the number of earthquake events per CA cell's is 1411 for the recorded earthquakes in a circular area of radius R = 80 km centred on the epicentre of the 26/7/2001 recorded earthquake (blue star) near Skyros Island, Greece during the period 1981-2005. The initial potential value of each cell is evaluated according to $V(i,j) = p \times M \times V_{th} \times char$, where p is a probability value strongly dependent to the recorded number of earthquake events at the cell (i,j) for the specific time period, M is the maximum recorded magnitude at this cell, V_{th} the threshold value of the potential and *char* is a parameter value related to the specific seismic characteristics of the area under test. The bottom plate shows the distribution of maximum earthquake magnitudes per cell of recorded earthquakes [60].

DNA Cellular Automata

Charilaos Mizas, Georgios Ch. Sirakoulis, Vasilios Mardiris, Ioannis Karafyllidis, Nicholas Glykos, and Raphael Sandaltzopoulos

The elementary cellular automata (CA) evolution rule can be extracted from a given number of CA evolution patterns and this can also be applied to the CAs that model DNA sequences. We map DNA to CA in such manner that sugar-phosphate backbone of a DNA molecule corresponds to the CA lattice and the organic bases to the CA cells. At each position of the lattice one of the four bases A (Adenine), C (Cytosine), T (Thymine) and G (Guanine) of the DNA molecule may be allocated, corresponding to the four possible states of the CA cell. We define as an *evolution event* a change in the state of one or more CA cells. Therefore, any point mutation (i.e. a mutation of a single nucleotide) is an evolution event and it corresponds to a cell state change. A time step in CA evolution is determined as the time interval between any pair of subsequent CA cell changes. So, the time flow is not uniform because mutations do not occur in regular intervals. Since CAs are deterministic computational models, their usefulness in modeling DNA evolution is proportional to the degree that DNA evolution is a deterministic procedure. Genetic Algorithms (GAs) are used to determine the rules of CA evolution that simulate the DNA evolution process. Linear evolution rules were considered and if DNA sequences of different evolution steps are available, the CA approach allows the determination of the underlying evolution rule(s). Once the evolution rules are deciphered, the reconstruction of the DNA sequence in any previous evolution step for which the exact sequence information was unknown becomes feasible. The paradigm described relies on the assumption that mutagenesis is governed by a nearest-neighbor dependent mechanism. Based on the satisfactory performance of the proposed model in the deliberately simplified example, the proposed approach could possibly offer a starting point for future attempts to understand the mechanisms that govern evolution.

C. Mizas · G.Ch. Sirakoulis · V. Mardiris · I. Karafyllidis
Department of Electrical and Computer Engineering, Democritus University of Thrace, Greece
e-mail: {cmizas,gsirak,vmardiris,ykar}@ee.duth.gr

N. Glykos · R. Sandaltzopoulos
Department of Molecular Biology and Genetics, Democritus University of Thrace, Greece
e-mail: {glykos,rmsandal}@mbg.duth.gr

© Springer International Publishing Switzerland 2016   127
A. Adamatzky and G.J. Martínez (eds.), *Art of Cellular Automata*,
Emergence, Complexity and Computation 20,
DOI: 10.1007/978-3-319-27270-2_21

DNA CA ©2008 Ch. Mizas, G.Ch. Sirakoulis, V. Mardiris, I. Karafyllidis, N. Glykos, and R. Sandaltzopoulos.

Evolution of a random DNA sequence modelled in one-dimensional CA. DNA bases that correspond to CA cells' states, are shown as A is blue, C is cyan, T is red, and G is yellow. Time arrow goes right. If the DNA sequence at various time steps of a line of evolution is available, it may be possible to determine the evolution rule (or rules) that triggered this evolution line. Once the evolution rule and the DNA sequence at present time are known, it may be possible to predict the next evolution event (or events) and, therefore, the DNA sequence at the next time step. For CAs with four-state, the number of possible rules is given by 4^{4^n}, where n is the number of cells in the neighbourhood CA rules are extended to 4^{4^3} or 4^{64}. The entire rule space of such CAs must be searched in order to find the putative CA evolution rules that govern the DNA sequence evolution. In the proposed CA model we assumed that DNA mutagenesis is influenced by the identity of the nucleotide to be mutated and the identity of the nucleotides in its vicinity. Based on this assumption, proper genetic algorithms were developed that efficiently extract the CA rule that governs sequence evolution [122].

Reversibility, Simulation and Dynamical Behaviour

Juan Carlos Seck Tuoh Mora, Norberto Hernandez Romero,
and Joselito Medina Marin

A cellular automaton (CA) is reversible if it repeats its configuration in a cycle. Reversible one-dimensional CA are studied as automorphisms of the shift dynamical system, and analyses using graph-theoretical approaches and with block permutations. Reversible CA are dynamical systems which conserve their initial information. This is why they pose a particular interest in mathematics, coding and cryptography.

Rule 110 CA are Turing-complete. The Rule 110 produces morphologically rich patterns composed of a periodic background (ether), on which a finite set of periodic structures (gliders) travel. The gliders collide and annihilate, produce new gliders or stationary localisations in result of the collisions. Rule 110 is a wonderful example of an apparently simple system with complex behaviour. This rule has been analysed using regular expressions, de Bruijn diagrams, and tilings. In particular, a block substitution system with three symbols is able to simulate the behaviour of Rule 110. The dynamics of Rule 110 can be reproduced by a set of production rules applied to blocks of symbols representing sequences of states of the same size, and the shape of the current blocks is useful for predicting the number of blocks in the next step. Another classic problem in CA is the specification of numerical tools to represent and study their dynamical behaviour. Mean field theory and basins of attraction have been commonly used; however, although the mean field theory gives the long-term estimates of density, it does not always give the adequate approximation for the step-by-step temporal behaviour. We present images related to the specification of reversible cellular automata by amalgamations of states and using memory. These are examples describing the dynamics of elementary CA by surface interpolation and the simulation of Rule 110 using a block substitution system.

J.C.S.T. Mora · N.H. Romero · J.M. Marin
Area Academica de Ingenieria, Universidad Autonoma del Estado de Hidalgo,
Carr Pachuca-Tulancingo Km 4.5, Col. Carboneras, Pachuca Hidalgo 42184, Mexico
e-mail: {jseck,nhromero,jmedina}@uaeh.edu.mx

© Springer International Publishing Switzerland 2016
A. Adamatzky and G.J. Martínez (eds.), *Art of Cellular Automata,*
Emergence, Complexity and Computation 20,
DOI: 10.1007/978-3-319-27270-2_22

Reversible one-dimensional CA of 9 states with neighbourhood size 2 and Welch indices $L = 9$ and $R = 1$. ©2014 Juan Carlos Seck-Tuoh-Mora.

Reversible CA conserve the information presented in their initial configuration. This is because these automata perform block permutations and shifts in every step. In one dimension, a reversible CA with s cell-states is equivalent to a full shift of s symbols. Symbolic dynamics operations can be applied over reversible automata in order to obtain and analyse different kinds of behaviours. In the case of a Welch index 1, the evolution rule of a reversible CA can be randomly defined by amalgamations of adequate permutations of states, obtaining in invertible dynamics as the one presented in the figure, characterised by cyclic patterns protected by quiescent barriers [142].

Reversible one-dimensional CA with 6 cell-states with neighbourhood size 2 in both invertible rules and Welch indices $L = 2$ and $R = 3$. ©2014 Juan Carlos Seck-Tuoh-Mora.

Reversible CA can be defined with a unitary Welch index, or both Welch indices different from 1. In any case, the product of Welch indices is equal to the number of de Bruijn blocks, or blocks with s^{2r} states; where s is the number of states and r is the neighborhood radius. For reversible automata $r = 1/2$ or neighbourhood size 2, the Welch indices product must be equal to the number of states. Even when both invertible rules are defined with $r = 1/2$, there are reversible automata able to generate interesting evolution patterns, as the one depicted in the figure of CA with 200 cells, evolving in 200 time steps [141].

Reversible one-dimensional CA with 8 **cell-states with neighbourhood size** 2 **in both invertible rules and Welch indices** $L = 2$ **and** $R = 4$. ©2014 Juan Carlos Seck-Tuoh-Mora.

Reversible CA with more states are able to produce more complicated behaviours, even when a neighbourhood size 2 is defining both invertible rules. The characterisation of reversible cellular automata by means of block permutations using the classical Cantor topology provides a basis for analysing its dynamical behaviour. With this, we obtain useful information about their transitive behaviour, in order to know if the automaton is topologically transitive or mixing [143].

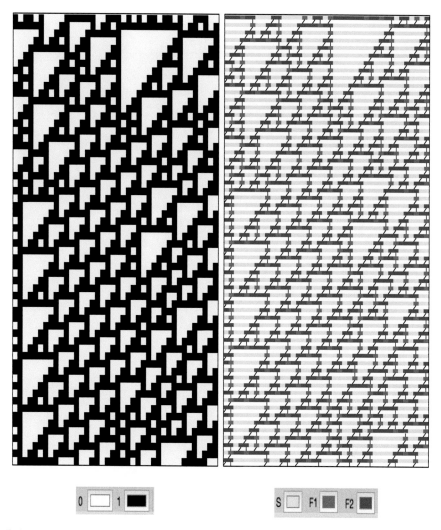

Rule 110 **represented as a block substitution system.** ©2014 Juan Carlos Seck-Tuoh-Mora.
Rule 110 can be represented as a block substitution system of three symbols: F_2 representing the
sequence 01, F_1 for state 1 iff it is on the left of 0 or F_2, and S_m for sequences of length m of the
same state between F_α, for $\alpha \in \{1,2\}$. With these symbols, the production rules reproducing Rule
110 dynamics are: $S_m F_2 \to S_{m-1} F_2 F_1$, $S_m F_1 \to S_{m-1} F_2$ and $S_m F_{\alpha_1} F_{\alpha_2} \dots F_{\alpha_p} \to S_{m-1} F_2 S_{q-2} F_1$
where $q = \sum \alpha_i$. The simulation is given by reordering blocks in every step when symbols F_α are
concatenated. In some cases, the substitution system conserves the information of the original
sequence, although the number of blocks may vary in this process, illustrating the dynamics of
Rule 110 as a combination of block mappings and re-orderings [145].

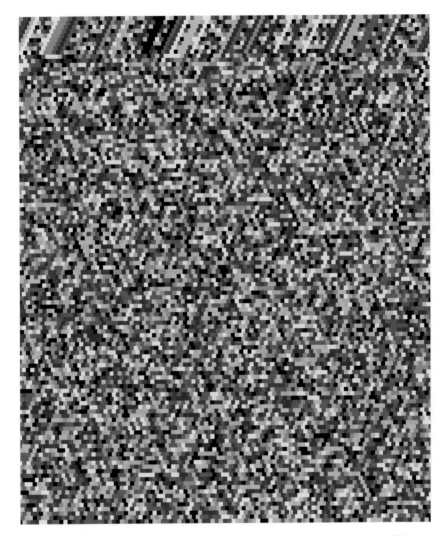

Reversible one-dimensional CA with 9 **cell-states, neighbourhood size** 2 **and invertible memory of size** 9. ©2014 Juan Carlos Seck-Tuoh-Mora.

We use a reversible automaton of 9 states and Welch indices $L = 9$ and $R = 1$. The evolution rule is complemented with an invertible memory function which takes into account the current generation and the previous 8 states of each cell. This memory is defined in such a way that the global evolution of the automaton is still reversible. The first nine generations are produced by the application of the classic reversible rule. From the ninth generation, however, the memory produces a more complicated dynamics, but still remains invertible [144].

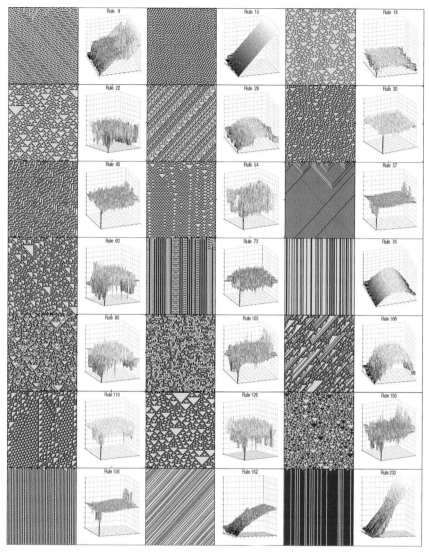

Surface interpolation for representing the dynamics of elementary CA ©2014 Juan Carlos Seck-Tuoh-Mora.

Mean field theory and basins of attraction are limited to explore the complex dynamics of elementary CA. The interpolation techniques can be applied to represent and classify the dynamics of elementary CA. The picture shows different interpolation surfaces obtained for elementary automata with hundred cells. A rich family of surfaces emerges in this analysis [146].

Aesthetics and Randomness in Cellular Automata

Nazim Fatès

We propose two images obtained with an asynchronous and a stochastic cellular automaton (CA). Deterministic cellular automata are now well-studied models and even if there is still so much to understand, their main properties are now largely explored. By contrast, the universe of asynchronous and stochastic is mainly a *terra incognita*. Only a few islands of this vast continent have been discovered so far. The two examples below present space-time diagrams of one-dimensional cellular automata with nearest-neighbour interaction. The cells are arranged in a ring, that is, the right neighbour of the rightmost cell is the leftmost cell, and *vice versa*; in formal words, indices are taken in $\mathbb{Z}/n\mathbb{Z}$, where n is the number of cells. The space-time diagrams are obtained with the FiatLux software[1]. Time goes from bottom to top: the successive states of the system are stacked one on the other.

N. Fatès
Inria Nancy – Grand Est
615 rue du Jardin Botanique, 54 600 Villers-lès-Nancy, France
e-mail: nazim.fates@inria.fr

[1] http://fiatlux.loria.fr

© Springer International Publishing Switzerland 2016
A. Adamatzky and G.J. Martínez (eds.), *Art of Cellular Automata*,
Emergence, Complexity and Computation 20,
DOI: 10.1007/978-3-319-27270-2_23

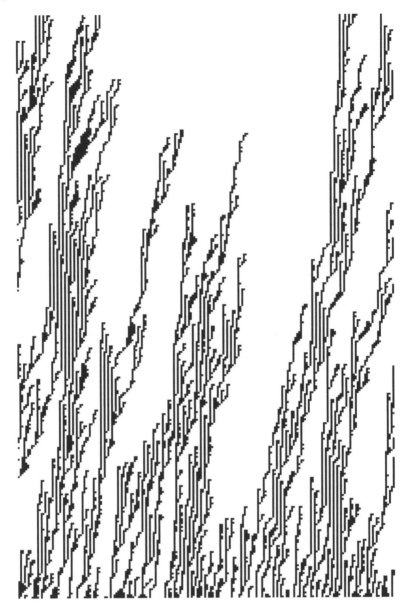

Aquatic plants: space-time diagram of the α-asynchronous elementary CA 148 =CFG. The evolution shows a particular evolution for $n = 100$ and $\alpha = 0.4$. Blue and white cells represent states 1 and 0. ©2015 Nazim Fatès.

An elementary CA is a one-dimensional binary CA with nearest-neighbour communication between cells. One of the simplest way of making this model asynchronous is to use α-asynchrony. With this updating scheme, a cell updates with probability α and keeps its state with probability $1 - \alpha$. The figure presents elementary CA Rule 148. It can alternatively be coded as rule CFG with the notation that lists the active transitions, see Ref. [51]. The behaviour of this rule is known for fully asynchronous updating [51], that is, a single cell is selected randomly at each time step, but it is still an open problem to determine its properties with α-asynchronous updating [50].

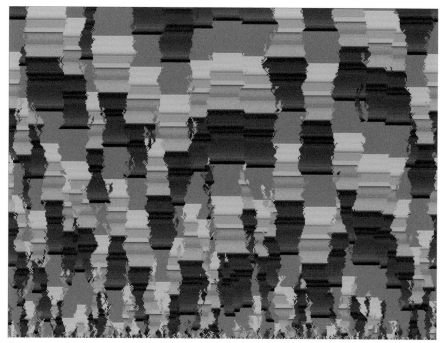

An *art nouveau* rug: space-time diagram of a stochastic CA designed for solving the global synchronisation problem. The setting is: $k = 130$, $n = 400$. Each state is associated with a colour drawn in palette arbitrarily constructed. ©2015 Nazim Fatès.

The global synchronisation problem is to make a CA blink at the same pace, whatever the initial condition. Genetic algorithms provide very good solutions to this problem [126]. However, it was recently shown that obtaining an *exact* solution of the problem with a deterministic rule is a difficult task [49]. By contrast, the problem is very easy to solve with a stochastic rule, and for any number of states. Assume that we have k states, at each time, each cell simply decides randomly and uniformly to either keep its own state or to copy the state of its left or right neighbour. Then, it updates its state by adding 1 to the state that was copied and making a modulo-k operation: $f(x,y,z) = \left(\mathcal{U}(x,y,z) + 1\right) \bmod k$, where $\mathcal{U}(x,y,z)$ selects of the three variables with uniform probability. As shown above, starting from a random initial condition where each cell selects one of 130 states independently and uniformly, the evolution of the rule make large portions of space synchronise progressively. Ultimately, the system should evolve to a uniform configuration and will then continue to "blink" for ever. Since the boundaries between homogeneous regions perform a non-biased random walk, the average time of synchronisation scales quadratically with the ring size n. It is an open problem to know if there is a rule whose average synchronisation time would scale linearly with n [49].

Cellular Automata with Memory

Ramon Alonso-Sanz

In conventional discrete dynamical systems, the new configuration depends solely on the configuration at the preceding time step. This contribution considers an extension to the standard framework of dynamical systems by taking into consideration past history in a simple way: the mapping defining the transition rule of the system remains unaltered, but it is applied to a certain summary of past states. This kind of embedded memory implementation, of straightforward computer codification, allows for an easy systematic study of the effect of memory in discrete dynamical systems, and may inspire some useful ideas in using discrete systems with memory (DSM) as a tool for modeling non-Markovian phenomena. Besides their potential applications, DSM have an aesthetic and mathematical interest on their own.

The contribution focuses on the study of systems discrete *par excellence*, i.e., with space, time and state variable being discrete. These discrete universes are known as cellular automata (CA) in their more structured forms, and Boolean networks (BN) in a more general way. Thus, the mappings which define the rules of CA or BN are not formally altered when implementing embedded memory, but they are applied to cells or nodes that exhibit trait states computed as a function of their own previous states. So to say, cells or nodes canalize memory to the mapping. An interactive implementation of the simplest types of CA with memory is accessible in the Wolfram Demonstrations Project[1]. Automata on networks and on proximity graphs, together with structurally dynamic cellular automata, have been also studied with memory.

Systems that remain discrete in space and time, but not in the state variable have been also scrutinised with memory. In particular, spatial games and maps in the complex plane. Both the Prisoner's Dilemma and the Battle of the Sexes games have been studied with memory in CA-like implementations. As a general rule, memory boosts cooperation in the former and coordination in the latter. Interesting variations of the dynamics of maps on the complex plane have been scouted dealing with the quadratic, cubic and Newtow-Raphson maps with memory in the complex plane [15, 16, 17]. In this scenario, further studies taking advantage of advanced visualisation techniques will likely provide appealing 2D patterns, generalising those achieved with the conventional dynamics without memory at the 80-90 decades of past century.

R. Alonso-Sanz
Technical University of Madrid, Spain
e-mail: ramon.alonso@upm.es

[1] http://demonstrations.wolfram.com/OneDimensionalCellularAutomataWithMemory

© Springer International Publishing Switzerland 2016
A. Adamatzky and G.J. Martínez (eds.), *Art of Cellular Automata*,
Emergence, Complexity and Computation 20,
DOI: 10.1007/978-3-319-27270-2_24

142

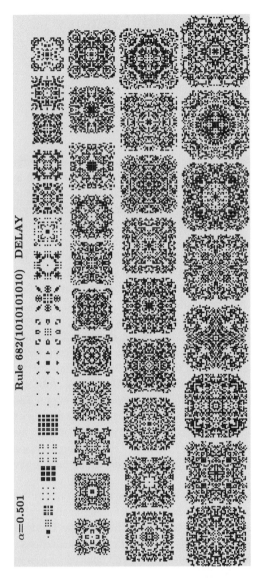

The 2D parity rule with very low memory from a single active cell. ©2011 Ramon Alonso-Sanz. In conventional discrete CA, the new configuration depends solely on the configuration at the preceding time step. In CA with memory, the mapping defining the transition rule of the system remains unaltered, but it is applied to cells with trait states computed as a certain summary of their past states. Here as a mean average with a geometrically discounted mechanism of factor α. In general, average memory of past states exerts an inertial effect which in the figure here restrains the highly chaotic dynamics of the parity rule even under the low memory charge induced by $\alpha = 0.501$ [14].

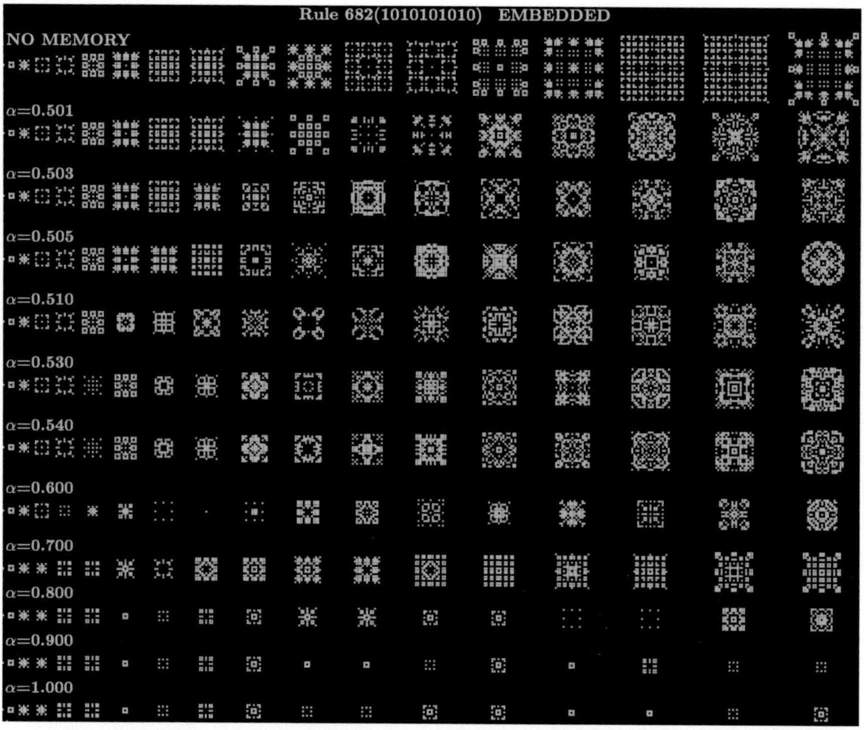

Dynamics of the 2D reversible parity rule with memory from a single active cell. ©2003 Ramon Alonso-Sanz. Second-order in time reversible rules may be implemented with memory of factor α as indicated in Fig. . In the figure here, the parity rule is endowed with increasingly memory charge tuned by the memory factor α [12].

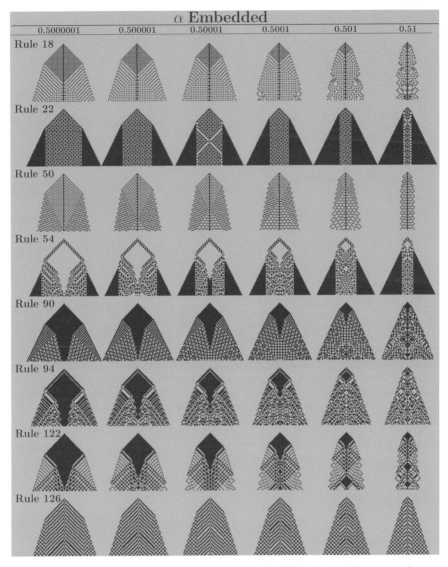

Fig. 1 Dynamics of 1D, reversible, legal elementary CA (ECA) rules with memory from a single active cell. ©2003 Ramon Alonso-Sanz. In two-state CA, α-memory is only effective if $\alpha > 0.5$, thus in this figure the memory charges are very low [18].

Dynamics of the 1D parity rule with ECA rules as memory from a single active cell. ©2006
Ramon Alonso-Sanz. Elementary CA rules may actuate as memory of length three. The 1D parity
rule (ECA 150) is endowed with ECA rules as memory labelled + in the figure here [19].

Dynamics of the 2D parity rule with ECA rules as memory from a single active cell. ©2006
Ramon Alonso-Sanz. The 2D parity rule endowed with ECA rules as memory labelled '+' in the
figure here. The dynamics starts from a single active cell, but it is shown in the picture here from
$T = 4$ [19].

R(14, 9) $\alpha = 0.60$ R(14, 9) $\alpha = 0.65$

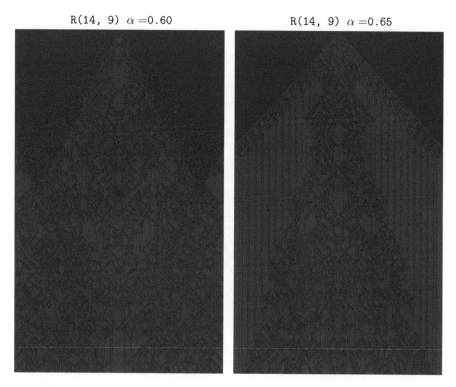

Actin CA with two low α-memory charges from a single active cell. The spatio-temporal patterns of the one-dimensional actin CA [10] may become highly intricate when endowed with memory. This is so even starting from a single active cell as shown in the above figures.

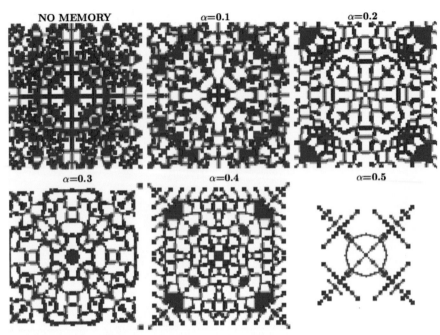

Patterns at $T=200$ in the spatial Prisoner's Dilemma starting from a single defector in the center of a lattice of size n=101 . Memory of past decisions and payoffs restrains the advance of defection noted by red (two consecutive defections) and blue (defection after cooperation) cells. White cells cooperate in two consecutive time-steps, and the green ones cooperate after defection [13].

Turing Machines and Checkerboards

Paul Rendell

The complexity of the behaviour that some simple cellular automata (CA) can exhibit is clearly shown by a demonstration of universal computation. The ability to perform any calculation that a Turing machine can make. The construction of such a machine necessarily requires areas covered by repeating patterns for storage of data and programs as well as areas of variation encoding information on a more direct way producing an uneven texture when viewed at a distance. These textures together with the inherent symmetry of the design lead naturally to artistically pleasing images.

The full universal Turing Machine constructed in Conway's Game of Life has a finite storage area for the universal Turing machine program made up of an array of 13 by 8 memory cells. It has unlimited storage for the specific Turing machine description and its data provided by two stacks which grow faster than the machine can push data onto them.

The two pictures of the full universal Turing machine in the Game of Life shown below produce interesting images for different reasons.

The first picture is appealing because of the tape construction wings which form the bulk of the image. These wings move and generate a stream of moving patterns called gliders. It is the debris from collisions between gliders from opposite directions which creates the components of the stack. The wings are a collections of identical moving patterns called rakes each of which produces one glider every 360 generations and move forward one square every two generations so the next glider is in the correct place to contribute to the next stack cell. However the rakes are not evenly spaced as it is their relative position that generates the specific stream of gliders required for the construction. A very simple algorithm was used to place the rakes therefore they are not packed as closely as possible resulting in a more interesting texture than would have resulted from a more optimal packing.

P. Rendell
Unconventional Computing Centre, University of the West of England, Bristol, UK
e-mail: paul@rendell-attic.org

© Springer International Publishing Switzerland 2016
A. Adamatzky and G.J. Martínez (eds.), *Art of Cellular Automata*,
Emergence, Complexity and Computation 20,
DOI: 10.1007/978-3-319-27270-2_25

The second picture shows traces of the paths of gliders during the initial phase where data for the stack is still being loaded as the stacks are being built. It thus shows the varied texture of the stack cells being constructed to the top left and bottom right. In the middle and top right are the regular patterns of the parts of the machine which make more sense when linked up by the tracks of gliders going between the components like delicate threads.

Holstein is a little known cellular automata with rule B35678/S4678. It uses the same neighbourhood of 8 cells as Conway's Game of Life however a cell is born if it has 3, 5, 6, 7 or 8 neighbours and survives if it has 4, 6, 7 or 8 neighbours. This rule is symmetrical in that a pattern of live cells in a background of dead cells behaves exactly the same as the same pattern in dead cells in a background of live cells. The result is generally quite boring as most patterns just shrink to nothing. There are a few small stable patterns and small oscillators and three large complex gliders have been found [41].

The checkerboard patterns where investigated as they frustrate the shrinking behaviour. In a closed universe where the top of the visible pattern is connected to the bottom and the sides connected to each other the checkerboard pattern is totally symmetrical the islands of one state are balanced by the islands of the other. The result of introducing a minimum of asymmetry to this initial pattern is quite astonishing. An amazing degree of symmetry is preserved throughout the pattern and the asymmetry is distributed across the universe.

With a very small number of exceptions the outcome is similar. The checkerboard pattern collapses to 4 blocks with stable boundaries round the universe in one direction and moving boundaries in the other. The moving boundaries move back and forth until two touch and then the pattern collapse again to a single band round the universe. My ongoing investigation is into why it takes so long for the moving boundaries to touch. The pictures below illustrate the very high degree of symmetry.

Full Universal Turing Machine in Conway's Game of Life. ©2015 Paul Rendell.
Two stack construction wings, top left and bottom right make the Turing Tape for the Turing machine in the middle. The main part of the machine is the array of memory cells containing the Universal Turing Machine Program. The data for the tape is provided by gliders just visible as a diagonal line of dots from the bottom left [134].

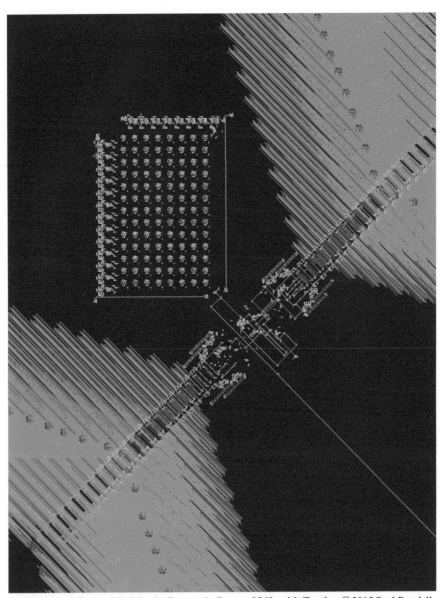

Full Universal Turing Machine in Conway's Game of Life with Tracks. ©2015 Paul Rendell. The original position of live cells are green, grey lines show where live cells have been and the position of live cells after 9000 generations is shown in black. It shows data being loaded from the line of gliders from the bottom left and pushed onto the bottom right stack as well as stack cell construction [134].

Holstein Checkerboard 1. ©2015 Paul Rendell.

Holstein is a two state two dimensional CA with rules B3578/S4678. These rules are symmetrical in that a pattern of one state in a background of the other state gives the same result regardless of which state is which. An interesting feature of this CA is that curved boundaries tend to straighten at a speed which is on average inversely proportional to the radius of curvature. This image was generated in a closed universe of 2240 × 2240 cells set up in an 16 × 16 checkerboard pattern. The symmetry confounds the tenancy of curved boundaries to straighten. The pure checkerboard pattern settles into an oscillating pattern. A single cell of asymmetry was added to create this pattern after 3500 generations. After the first generation the number of cells in each state is identical and remains so for ever.

Holstein Checkerboard 2. ©2015 Paul Rendell.
The same Holstein Checkerboard pattern as the previous page run for 35000 generations. Run further the blobs on the right shrink to the centre line and disappear leaving 4 blocks of alternate state (the left and right edges should be considered to be joined). This pattern persists with the edges moving left and right until they touch and the pattern collapses into horizontal bands each with the same number of cells. For a universe of this size this is estimated to occur some time after 10^{11} generations. The diamond patterns of dots are period two oscillators created in an initial phase when the straight edges of the original checkerboard pattern are replaced by oscillating waves. It is astonishing how many survive the change from 16×16 to 2×2 blocks.

Aperiodicity and Reversibilty

Katsunobu Imai

We show cellular automata (CA) configurations obtained in our studies of space-time dynamics on aperiodic tilings and reversible self-reproducing CA. The families of CA considered have almost nothing in common but their unconventionality. In 2012 Goucher found a glider propagating CA on Penrose tilings [63]. His work was inspired by the study of the variants of the Game of Life on Penrose tilings conducted by Owens and Stepney [127, 128]. CAs on quasi-periodic tilings are frequently referred to as a framework of simulation of signals in automaton ensembles because Penrose tilings seems to be more homogeneous in terms of signal propagation. First figure presents spiral generated on the Penrose tiling. Spirals generated by cyclic CA on several quasi-periodic tilings have been studied by Reiter [132]. We propose a simpler version. A CA of family $S/B/C$ is a semi-totalistic CA, where S (B) represents the number of live cells in the neighborhood of a cell necessary for the cell's survival (birth). $C \geq 3$ is an integer called generations parameter which is equal to the size of its states set. When an alive cell can not survive, its state changes from 1 to 0 by way of $C-2$ refractory states, i.e., $1 \to 2 \to \cdots \to C-1 \to 0$. Typically, spirals generated by such simple generations CA on a square grid have an "artificial" square looks (e.g. 2/234/5). On Penrose tilings, thanks to the quasi-periodic properties of their cellular space, it is possible to get more "smooth" patterns of wave propagation including spiral patterns generated from a small seed of two live cells [78].

The spirals generated look 'circular' but, of course, they are not truly homogeneous. The outer shape of a spiral finally converges to the decagonal shape. In general, the shape of a simple diffusion on a quasi-periodic tilings projected by its n-dimensional dual space finally converges to the regular $2n$-sided polygonal shape (Penrose tilings are projected by a 5-dimensional dual space). The speed of the convergence depends on exact type of tiling and cell neighbourhood. When the Penrose tilings are employed, the convergence speed is slower in the combination of rhombus

K. Imai
Hiroshima University, 1-4-1, Kagamiyama, Higashi-Hiroshima, 739-8527, Japan
e-mail: imai@hirosihma-u.ac.jp

© Springer International Publishing Switzerland 2016
A. Adamatzky and G.J. Martínez (eds.), *Art of Cellular Automata,*
Emergence, Complexity and Computation 20,
DOI: 10.1007/978-3-319-27270-2_26

tiles and Moore neighbourhood than the other three cases (kite-dart and von Neumann neighbourhood, kite-dart and Moore, and rhombus and von Neumann). The
slow convergence derives from the complex formation of the outer shape. Even the
evolution of a simple diffusion rule, its implicit signal collisions are quite complicated in the case of rhombus tiles and Moore neighbourhood. Although a four-state
(weak) universal in rhombus Penrose tile CA is known so far, a simpler Penrose tile
CA may have more computational complexity than apparently seen.

Second figure is a configuration of a reversible self-reproducing CA. It is realised by modifying the Langton's 8-state self-reproducing CA. We shown that such
a self-reproducing system can be embedded also in a reversible cellular space. Two
kinds of objects — worms and loops — are introduced and their self-reproductions
are performed by a self-inspective manner. The automaton's global transition is bijective and supports a large variety of self-reproducing behaviours. The reversible
self-reproducing automaton is designed using the framework of partitioned cellular
automaton: a cell is partitioned into seven parts and each part has nine states [79].
The self-reproducing process is performed reversibly with no dissipation. Self-
reproduction with dissipation contradicts widely accepted belief that all organisms
are dissipative structures. Physically speaking they are indeed. However, at the abstract level of automata it is still possible to produce a reversible artificial life.

Wave propagations in cellular automaton $/23/5$ on a kite and Dart Penrose tiling. ©2014 Katsunobu Imai.

A CA generations family $S/B/C$ is a semi-totalistic CA, where S (B) represents the number of live cells in the neighbourhood of a central cell to survive (to be born) respectively. $C \geq 3$ is an integer called generations parameter which is equal to the size of its states set. When an alive cell can not survive, its state changes from 1 to 0 by way of $C - 2$ refractory states, i.e., $1 \to 2 \to \cdots \to C - 1 \to 0$. Normally, spirals generated by such simple generations CA on a square grid have quite "artificial" looks (e.g. 2/234/5). But in the case of Penrose tilings, thanks to the quasi-periodic properties of the cellular space, it is possible to get more homogeneous patterns of wave propagation including spiral patterns just starting from a small seed (two live cells) [78]. See also https://www.youtube.com/watch?v=o_5aHvkm3d4 for the animation.

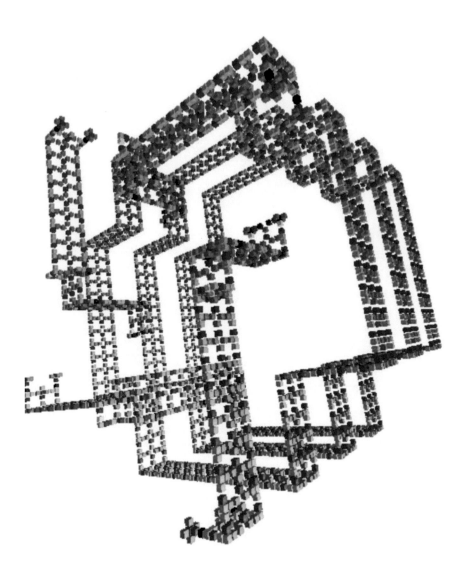

Three-dimensional reversible self-reproducing CA. ©2014 Katsunobu Imai.
This is an example configuration of a reversible self-reproducing CA. It is realised by extending the Langton's 8-state self-reproducing CA. Two kinds of objects (*worm* and *loop*) are introduced and their self-reproductions are performed by a self-inspective manner. Even its global transition is bijective, it supports a variety of self-reproducing behavioural patterns. It is designed using the framework of partitioned CA. Each cell is partitioned into seven parts and each part has nine states [79].

Painting with Cellular Automata

Danuta Makowiec

Images created by Nature fascinate us, and always have. People try to reproduce them in different forms, according to their own skills and/or personal feelings. Artists, for example, use sounds, colours, stones to express their fascination with what they see or what they sense. Scientists, on the other hand, demonstrate their fascination through their strong desire to understand natural phenomena. Discoveries being made day by day in physiology labs may revise many of our conceptions, including the concept of computation. However, painting was and still is a commonly used method of describing new observations, because painting involves understanding. The cellular automata system is a mathematical concept that paints pictures of certain effects. Often it is the case that the results that cellular automata provide are understood intuitively rather than being mathematically well proven. Below we present a 'work of art' produced by automata.

Self-sustained oscillators appear to be ubiquitous in nature. These are active systems that contain some internal source of energy, which is then transformed into oscillatory dependence; see the book by Pikovsky, Rosenblum and Kurths to learn of many examples [130]. Such oscillating entities, when interacting with one another, may lead to the emergence of states with regular patterns. For example, biologists have observed that excited myocytes — cells which are the main substance of the tissue of the heart — gather in patterns in the shape of spirals or radiating circles. Note that these natural computations are space distributed and dynamically changed due to the cascade of biochemical processes involved. To make sense of the complex interactions and dependencies, science must combine perspectives at the component level and the system level. Such perspectives lend themselves to the cellular automata approach, because cellular automata modelling is pragmatic. Avoiding molecular details, it relies on a phenomenological description of a complex picture of the cell biochemistry to produce patterns similar to those obtained by physiologists observing tissue properties.

D. Makowiec
Institute of Theoretical Physics and Astrophysics, University of Gdansk, Poland
e-mail: fizdm@univ.gda.pl

© Springer International Publishing Switzerland 2016
A. Adamatzky and G.J. Martínez (eds.), *Art of Cellular Automata*,
Emergence, Complexity and Computation 20,
DOI: 10.1007/978-3-319-27270-2_27

159

Cyclic cellular automata (CA) modelling of natural heart pacemaker. ©2015 Danuta Makowiec.

Every cell progresses through three consecutive states: firing, refractory, and active. In each state, a cell spends a fixed number of time steps: n_f, n_r, n_a, respectively. This constitutes a cellular cycle of length $n_f + n_r + n_a$. This intrinsic cellular cycle is changed by neighbouring cells. An active cell fires if the number of neighbours firing is greater than $\mathbf{F}$, $0 \leq \mathbf{F} \leq 8$. The clock of a refractory cell is turned back half an actual advance in refractory state if the number of neighbours firing is greater than $\mathbf{R}$, $0 \leq \mathbf{R} \leq 8$; otherwise the cell follows the intrinsic cycle. The neighbourhood of each cell consists of $\mathbf{n}$ randomly chosen neighbours $0 \leq \mathbf{n} \leq 8$ in the Moore neighbourhood on a square lattice. The configuration when all the cells are in the same phase of the intrinsic cycle is invariant and is an attractor. However, at certain parameter settings, the automata produce periodic and stable signals due to specific self-organised geometric constraints, sometimes in the shape of a spiral, but often reminiscent of radiating or collapsing circles produced by sustained engines. The examples of such relations given above were obtained when starting from a random initial state. Firing cells are red, refractory white, and active green; $n_f = 9, n_r = 11, n_a = 19$; a rule is defined by a triple $(\mathbf{n}, \mathbf{F}, \mathbf{R})$; $*$ refers to the conditions which best reveal the physiology of the natural human pacemaker [92].

Patterns in Cellular Automata

Harold V. McIntosh

The idea of a one dimensional cellular automaton (CA) is quite simple, and its evolution in time is ideal for a two dimensional presentation, as on a video screen. To
start with, a cell is a region, even a point, with differing forms, called states. For
convenience, these states are usually numbered with small integers beginning with
zero, rather than described. For the purposes of automata theory the nature of the
states does not matter, only their relation to one another, and the way they change
with time according to their environment. Since they are abstract, they can just as
well be represented by coloured dots on a video screen, which is what makes them
so dramatic when interpreted as an abstract artistic design.

Unfortunately the quantity of computation required to simulate an automaton
increases exponentially, and usually with a rather large exponent, with the size of
the parameters involved. Thus any increase in the number of parameters will only
aggravate a situation which is already rather difficult. If a significant practical application of one of these variants were to be discovered, no doubt the means would
be found to compute its properties. In the meantime the regular version of cellular
automata has enough unsolved problems to keep one occupied.

There are two visible tendencies. One is the approach of John Conway [59], or of
those who already have an automaton and want to find out what it does. The other
is the original approach of John von Neumann [163], to go ahead and construct the
automaton which one requires, regardless of the cost in cells and states. Given that a
Turing machine can be embedded in cellular automata in a fairly standard way, there
will always be some automaton which will realise a given calculation; typically such
a straightforward realisation will be neither aesthetic nor efficient.

Fundamental to finite automata is eventually [116] periodic evolution, prolonged
to the full number of states in the exceedingly rare event that their sequence is cyclic.
The number of "states" of a cellular automaton is the number of configurations (not
states per cell), providing an exponentially large bound relative to the automaton's

H.V. McIntosh
Departamento de Microcomputadoras, Universidad Autónoma de Puebla, Puebla, México
e-mail: mcintosh@unam.mx

© Springer International Publishing Switzerland 2016   161
A. Adamatzky and G.J. Martínez (eds.), *Art of Cellular Automata*,
Emergence, Complexity and Computation 20,
DOI: 10.1007/978-3-319-27270-2_28

length. In practice, many short cycles usually predominate over a few long ones, almost always reached through transients.

Longer automata admit longer cycles and longer transients too; the infinite limit may lack cycles. Cyclic boundary conditions locate behaviour repeating over a finite range, leaving truly aperiodic configurations for a separate study.

A great advantage of working with one dimensional CA is that their time evolution can be shown on a two dimensional chart, whereas the evolution of a two dimensional structure would require a third dimension. While not impossible to show, there is too much information involved to keep the presentation from becoming extremely cluttered. Even so, an infinite, or even a very long, line is hard to manage. A systematic study could easily start with short lines, folded around to form a ring, avoiding end cells whose rules of evolution would differ from the interior cells. For short enough rings, the complete evolution of all possible configurations can be calculated.

Fortunately, for one dimensional automata, a diagrammatic technique which lies at the heart of shift register theory saves the situation; the diagrams are called de Bruijn diagrams, but they are just simple graphs showing the possible ways in which different neighbourhoods can overlap [113].

In principle, such a diagram could be extended to automata of higher dimensions, but a problem arises from selecting partial neighbourhoods that will join to form full neighbourhoods in all directions. The straightforward approach of building up strips of successively higher dimension runs afoul of Post's correspondence principle when arbitrary intermediate strips have to be matched to form the strips of the next higher dimension. If only periodic solutions are required, the problem is still soluble, but again the conflict between large systems and unbounded systems arises, tending to leave the generic properties of aperiodic systems undecidable [116].

On the other hand, when statistical methods began to be applied to the evolution of CA, it was understood that there was a difference between periodic configurations for the automaton and periodicities in the statistical behaviour of the automaton. The long time behaviour of a simple configurations is one thing; it may repeat sooner or later exhibiting periodic behaviour, although if the evolution consists of a never ending transient, it may still have convergent statistical properties. But it is another thing to examine statistical behaviour averaged over all configurations, or all the members of a class of configurations. It has generally been expected that those averages would converge, and there was even an article published arguing to that effect. Shortly thereafter Hugues Chaté and Paul Manneville [28] discovered some automata which, although they did not exactly follow the return map of iteration theory, did not tend toward a long term average either. The most prominent of them, discovered shortly after their original announcement by Jan Hemmingsson, was only three dimensional and followed a period of three in the density [114].

Cycle diagrams as galaxies patterns. ©2009 Harold V. McIntosh. There are two ways to obtain the cycles for a given automaton. The first is to enumerate all the rings of the desired length, and follow up the evolution of each. In doing so various shortcuts can be taken, such as generating the configurations in Gray code order so that only a single cell changes state from one to the next. Still lifes can be detected very quickly this way. Numerical comparison of successive generations means that whenever the new generation is smaller, it will already have been examined and need not be pursued further. The second way is more systematic and is worth the bookkeeping effort involved. A graph whose links are determined by evolution is prepared, following which a path enumerating procedure is followed to locate all the loops, whose lengths will give the periods of all the cycles of that length. Cycles of length up to ten can be obtained easily, twenty with effort, but passing thirty requires dedication; for binary automata it is slightly easier, increasingly more difficult for others [116]. This plot displays a cycle diagram with a cycle attractor of length six for an automaton $(4, h)$ evolution rule FA9098C2.

Generic de Bruijn diagram. ©2009 Harold V. McIntosh. Looking in another direction, the cells of one dimensional automata, by definition, form linear chains. But the neighbourhoods themselves form another kind of chain, wherein they necessarily overlap. The result is an arrangement which has a very close connection with shift register theory, which is a systematic study of the properties of overlapping strings and how the overlaps happen to come about. In particular, the form of graphical representation known as the de Bruijn diagram [113] enters into many discussions, and can be used to organise a major portion of the theory. This plot display a generic de Bruijn diagram for elementary CA in four generations without any displacement.

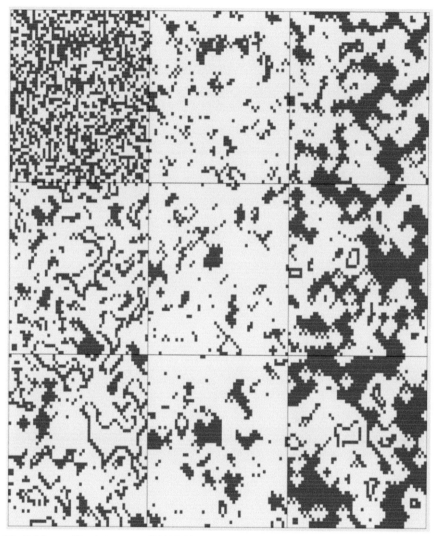

Chaté-Manneville Automata. ©1999 Harold V. McIntosh. The density of states in Chaté-Manneville automata [28] runs through cycles rather than approaching an equilibrium value. One of simplest is the binary totalistic rule 33 in three dimensions with von Neumann neighbourhoods. This stages of evolution displays a plane cross section of the evolution of a random initial configuration for nine generations, non-trivial collective behavior can be distinguished because different densities repeat exactly every three steps.

Gliders in One-Dimensional Cellular Automata

Genaro J. Martínez

Gliders are non-trivial complex patterns emerging typically in complex cellular automata (CA). The most famous glider (mobile self-localizations, particles, waves) is the five-live cells glider moving diagonally in five steps in the two-dimensional CA Conway's Game of Life [59]. Gliders are abstractions of travelling localizations often found in living systems, and used to derive novel properties of objects in artificial life, complex systems, physical systems, and chemical reactions. Examples include, gliders in reaction-diffusion systems [3, 125], Penrose tilings [63], three-dimensional glider gun [2], gliders in hyperbolic spaces [98].

Gliders are found in one-, two- and three-dimensional automata, in complex cell state transition rules and chaotic rules. All space-time configurations of cellular automata (CA), shown in this chapter, use gliders in one dimension and are coded using regular expressions derived from the de Bruijn diagram and tiling theory. We use filters to eliminate the periodic background and to get a better view of gliders. The filters are useful tools for inspecting details of glider collision, especially when configurations are very large.

Pictures shown here are space-time configurations of elementary CA rule 110 [101]. In 2004 Cook demonstrated that the rule 110 CA is universal because it simulates a cyclic tag system. The cyclic tag system if programmed into an initial configuration of CA. The processing of just four values on the tape of this machine requires over three millions of cells [32, 165, 108, 31].

G.J. Martínez
Escuela Superior de Cómputo, Instituto Politécnico Nacional, México,
Unconventional Computing Centre, University of the West of England, Bristol, United Kingdom
e-mail: genaro.martinez@uwe.ac.uk

© Springer International Publishing Switzerland 2016
A. Adamatzky and G.J. Martínez (eds.), *Art of Cellular Automata*,
Emergence, Complexity and Computation 20,
DOI: 10.1007/978-3-319-27270-2_29

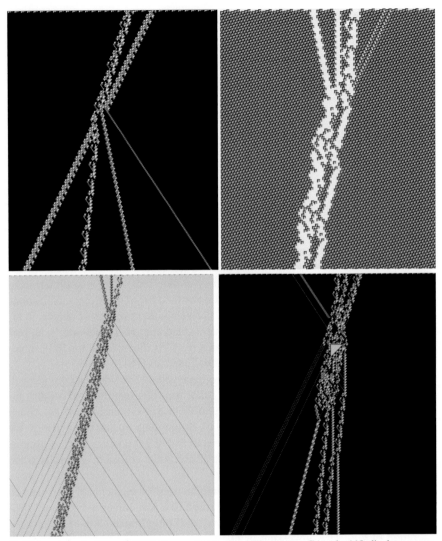

Collisions in rule 110. ©2010 Genaro J. Martínez. Elementary CA rule 110 displays a nontrivial behaviour first discovered by Wolfram in 1986 [163]. The rule is proved to be a universal by simulating a cyclic tag systems [32]. Rule 110 support 12 gliders and one glider gun [106, 115]. Top left pattern shows a collision between two gliders (G and $\bar{B}$ in Cook's nomenclature), splitting in four gliders ($\bar{B}$, F, D_2, and A^3). After the collision, one glider continues along its original trajectory while second splits into five gliders. Top right picutre displays a collision between five gliders (D_1, C_3, F, and $2B$), they produce a larger glider (H) in the result of the collision. Bottom left picture shows a triple collision (D_1, C_1, and $\bar{E}$) that results in formation of a glider gun. Bottom right space-time configuration show collision between gliders to get a big tile in rule 110; ten gliders (A, A^5, $2F$, B, and G) are necessary to yield a T_{30} tile [107].

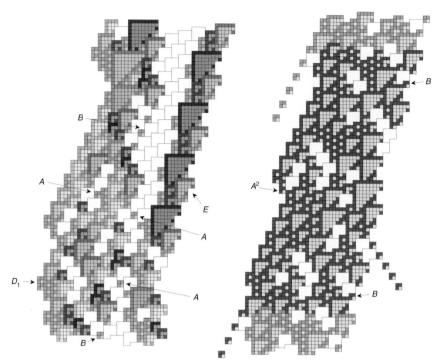

Tilings and Complex Dynamics ©2001 Genaro J. Martínez. The automaton rule 110 displays a diversity of gliders (mobile self-localizations, particles, waves). The rule 110 can be assembled just with polygons. This figure shows in full details what is the number of polygons, kind, interaction, and position to construct a *H* glider (largest glider in rule 110) and a glider gun [106].

Meta-Glider in Rule 110 ©2007 Genaro J. Martínez. A meta-glider is constructed with a lot of gliders colliding, these collisions are synchronised and repeated cyclically. These meta-gliders are very sensitive to any external perturbation and they can be destroyed in few steps. Elementary CA rule 110 supports a number of meta-glider constructions [109]. This simulation displays a concatenation of five gliders colliding in different times to get a new series of six gliders. To synchronise the whole set of collisions we use the phase codification of the gliders. The collisions between gliders can be also seen as solitonic collisions [103].

Gliders in CA with Memory ©2010 Genaro J. Martínez.

In CA with memory cells update their states depending on cumulative states of their neighbour-hoods [14]. A systematic analysis of CA with memory is done in [102]. In [110] we shown that a 'classical' chaotic rule — rule 126 — exhibits a complex behaviour when equipped with majority memory. CA with memory $\phi_{R126maj:4}$ display an amazing range of collisions between gliders. The space-time configuration shows evolution of the binary CA from one cell in state '1' during one thousand of generations. Colours represent different periodic backgrounds.

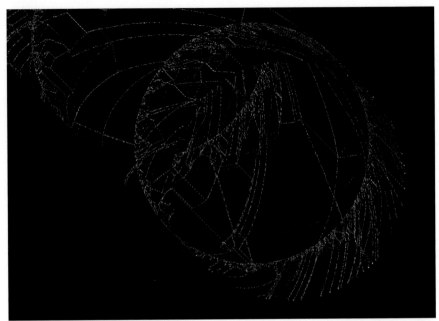

CA Collider ©2011 Genaro J. Martínez. Computations in CA are implemented from von Neumann's era till present, and a number of different designs have been produced [100] using signal interaction, glider collisions, tiling assembling, self-reproduction machines, or deriving formal languages. In [111, 105] we shown how to implement computation in one-dimensional CA using cyclotrons and virtual colliders. Our CA collider works with a finite set of cyclotrons. We design a finite state machine where nodes are meta-nodes than represent states of a set of gliders. A transition between two nodes of the CA collider is a result of cascaded collisions between gliders. This figure shows how a number of gliders travel a long of a cyclotron, and the history of all collisions and trajectories are projected in three dimensions. This simulation presents thousand of gliders in 20,000 cells space. All non-trivial patterns are represented just as dots. Details of any particular structures are not relevant just the kind of collision and its result. An interesting design relates several cyclotrons synchronised to simulate a cyclic tag system in rule 110 in a virtual collider [104].

Excitable Automata

Andrew Adamatzky

Cellular automata (CA) are computationally efficient and user-friendly tools for simulation of large-scale spatially extended locally connected systems. CA representation of reaction-diffusion and excitable systems is especially interesting because this allows us to effortlessly map already established massively-parallel architectures onto novel material base of chemical systems, and also design novel non-classical and nature-inspired computing architectures [9]. The examples of 'best practice' include cellular automata models of Belousov-Zhabotinsky reaction, [61, 99], chemical systems exhibiting Turing patterns [175, 172] precipitating systems [9], calcium wave dynamics [173], and chemical turbulence [72]. Reaction-diffusion modelling and simulation, particularly in a sense of chemical computation and development of wave-based chemical processors [9, 8], becomes a hot topic of computer science, physics and chemistry. A distinctive feature of the CA-based prototyping is that it is made on an intuitive, we can say interpretative rather then implementative, level, where automaton states are seen as chemical species and cell-state rules as quasi-chemical reactions [8]. That is we do not have to follow reaction-diffusion dynamic to simulate it in automata but instead we should map all possible models of cellular automata onto a space of quasi-chemical reactions. Here we present configurations generated by cellular automata rule set of which were discovered during our quest for minimal architecture of reaction-diffusion computers and the role of 'primitive' structures — gliders, self-localisations, particles – in shaping space-time mosaic of emergent computation [8]. The examples include, excitable CA where a cell can stay excited for more than one iteration, waves of excitation on automaton network derived from Delaunay triangulation, reaction-diffusion automaton where two chemical species facilitate each other propagation, excitable CA where boundaries of excitation interval change during the automaton evolution, and propagation of disturbance in actin polymer automata.

A. Adamatzky
Unconventional Computing Centre, University of the West of England, Bristol, UK
e-mail: andrew.adamatzky@uwe.ac.uk

© Springer International Publishing Switzerland 2016
A. Adamatzky and G.J. Martínez (eds.), *Art of Cellular Automata,*
Emergence, Complexity and Computation 20,
DOI: 10.1007/978-3-319-27270-2_30
173

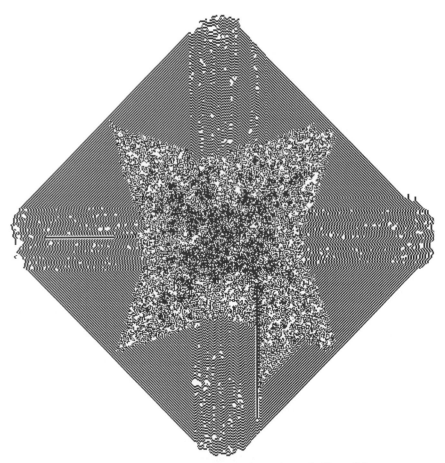

Astroid-shaped irregular excitation and refractoriness in excitable cellular automaton
$R(2448)$ **with retained excitation.** ©2015 Andrew Adamatzky.
Every cell of the automaton takes three states: resting, excited and refractory. A resting cell
excites if a number of excited neighbours belongs to interval $[\theta_1, \theta_2]$, $1 \leq \theta_1 \leq \theta_2 \leq 8$; the cell
remains resting otherwise. An excited cell remains excited if a number of excited neighbours
belongs to interval $[\delta_1, \delta_2]$, $1 \leq \delta_1 \leq \delta_2 \leq 8$; the excited cell takes refractory state otherwise. We
denote rules by tuples $R(\theta_1 \theta_2 \delta_1 \delta_2)$. Several modes of space-time activity dynamics are discovered
by analysing 1296 rules of retained excitation: not growing but persistent domains of activity,
domains with rectangular, octagonal and almost circular growth, amoeba-like growing patterns,
mobile and still localizations [4].

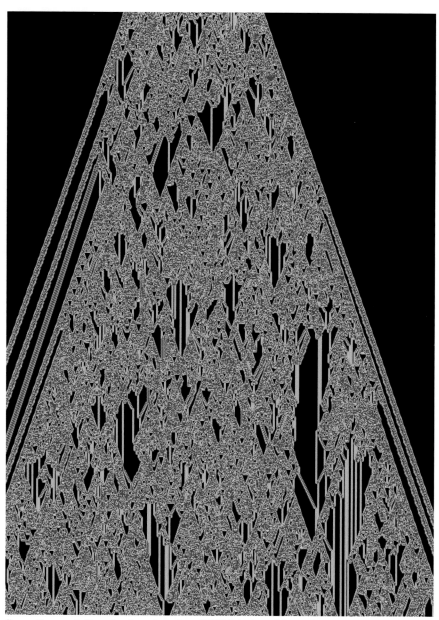

Space-time evolution of actin automaton. ©2015 Andrew Adamatzky.
Actin is a globular protein which forms filaments in the eukaryotic cytoskeleton. Its roles include structural support, contractile activity, and intracellular information processing. We model actin double helix as two chains of one-dimensional binary-state semi-totalistic automata. Each node of the actin automaton takes state '0' (resting) or '1' (excited) and updates its state in discrete time depending on its neighbour's states. Time evolves downwards. Red pixels show where complementary sites of both chains are in state '1', green where only one site is in state '1', black otherwise [10].

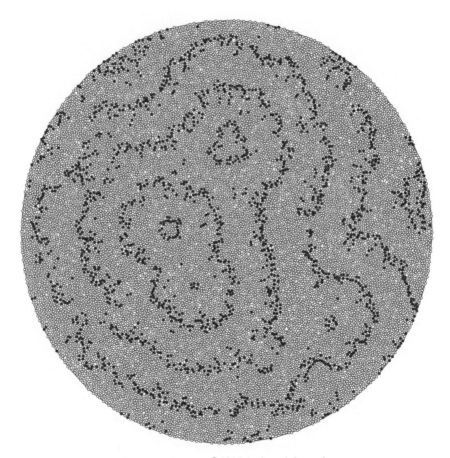

Excitatiton waves on Delaunay automata. ©2015 Andrew Adamatzky.

Given a planar finite set $\mathbf{V}$ the Delaunay triangulation $\mathscr{D}(\mathbf{V}) = \langle \mathbf{V}, \mathbf{E} \rangle$ is a graph subdividing the space onto triangles with vertices in $\mathbf{V}$ and edges in $\mathbf{E}$ where the circumcircle of any triangle contains no points of $\mathbf{V}$ other than its vertexes. Neighbours of a node $v \in \mathbf{V}$ are nodes from $\mathbf{V}$ connected with v by edges from $\mathbf{E}$. The Delaunay automaton is defined in the following way. A node $v \in \mathbf{V}$ is a finite state machine. Every node updates its state in discrete time depending on states of its neighbours. All nodes update their states simultaneously. Nodes can have different number of neighbours therefore we better use totalistic node-state update function, where a node updates its state depending on just the numbers of different node-states in its neighbourhood. We assign three states — resting (gray), excited (red) and refractory (blue) — to nodes of $\mathbf{V}$. A resting node excites depending on a number of excited neighbours. If a node is excited at time t the node takes refractory state at time step $t + 1$, independently on states of its neighbours. Transition from refractory to resting state is also unconditional. See details in [6].

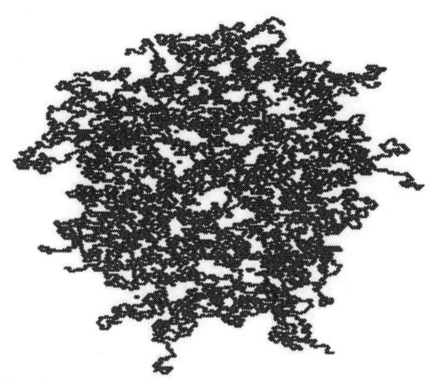

Bunch of worms in mutualistic automaton. ©2015 Andrew Adamatzky.
We simulate mutualistic relationships in a two-dimensional hexagonal cellular automaton. Every cell takes three states: 0, 1 and 2. States 1 and 2 represent species species '1' and species '2'. State 0 represents an 'empty space', or a substrate. Two processes must be simulated: propagation of species and survival of species. A cell of the hexagonal cellular automaton can be occupied exclusively by 'empty space', state 0, or by one of the species, states 1 or 2. A cell x in 'empty state' at time t ($x^t = 0$) becomes occupied by species 1 at time $t+1$ if it has more neighbours in state 1 than in state 2 ($\sigma_1^t > \sigma_2^t$, $\sigma_i^t = |\{y \in u(x) : y^t = i\}|$, $i \in \{1,2\}$) but enough neighbours in state 2 to support species 1 depending on them ($\sigma_2^t > \theta_{01}$), or it has equal amount of both species ($\sigma_1^t = \sigma_2^t$) but there are more species 2 to support species 1 than species 1 to support species 2 ($\sigma_2^t - \theta_{01} > \sigma_1^t - \theta_{02}$). Propagation of species 2 can be discussed similarly. Worms are quasi-one-dimensional propagating localizations, which usually consists of two parallel chains of non-quiescent states, one chain is formed by sites with species 1, another chain by sites with species 2. All worms observed in our experiments have only two growing tips [5].

Configurations of excitation intervals in cellular automata with dynamical excitation intervals ©2015 Andrew Adamatzky.

In excitable cellular automaton a cell takes three states: resting, excited and refractory. An excited cell takes refractory, and a refractory takes excited state unconditionally. In excitable automaton with dynamical excitation intervals a resting cell becomes excited if number of its excited neighbours belong to some interval. Boundaries of the excitation interval of a cell are also updated, at every of automaton evolution, depending on ratio of excited to refractory neighbours of the cell. Excitable cellular automata with dynamical excitation interval exhibit a wide range of space-time dynamics based on an interplay between propagating excitation patterns which modify excitability of the automaton cells. Such interactions leads to formation of standing domains of excitation, stationary waves and localised excitations [7]. In the snapshot of excitation intervals values of the upper boundary are encoded by colours as follows: 1 is white, 2 is green, 3 is yellow, 4 is blue, 5 is magenta, 6 is cyan, 7 is red, and 8 is black

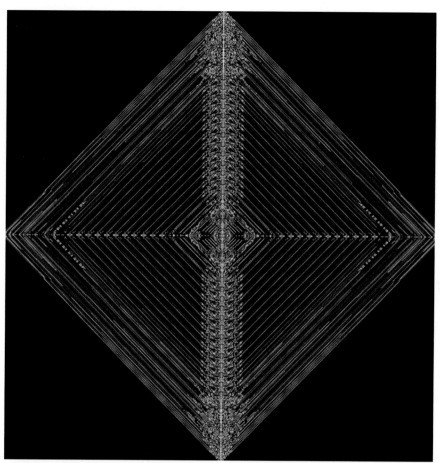

Travelling localisations in in cellular automata with dynamical excitation intervals ©2015
Andrew Adamatzky.
In excitable cellular automaton a cell takes three states: resting, excited and refractory. An excited
cell takes refractory, and a refractory takes excited state unconditionally. In excitable automaton
with dymamical excitation intervals a resting cell becomes excited if number of its excited neighbours belong to some interval. Boundaries of the excitation interval of a cell are also updated, at
every of automaton evolution, depending on ratio of excited to refractory neighbours of the cell [7].
In the snapshot of excitation intervals values of the lower boundary are encoded by colours as follows: 1 is white, 2 is green, 3 is yellow, 4 is blue, 5 is magenta, 6 is cyan, 7 is red, and 8 is black.

References

1. Achasova, S., Bandman, O., Markova, V., Piskunov, S.: Parallel Substitution Algorithm. Theory and Application. World Scientific, Singapore (1994)
2. Adamatzky, A.: Universal dynamical computation in multidimensional excitable lattices. International Journal of Theoretical Physics **37**(12), 3069–3108 (1998)
3. Adamatzky, A.: Computing in nonlinear media and automata collectives. Institute of Physics Publishing (2001)
4. Adamatzky, A.: Phenomenology of retained excitation. International Journal of Bifurcation and Chaos **17**(11), 3985–4014 (2007)
5. Adamatzky, A.: Localizations in cellular automata with mutualistic excitation rules. Chaos, Solitons & Fractals **40**(2), 981–1003 (2009)
6. Adamatzky, A.: Excitable delaunay triangulations. Kybernetes **40**(5/6), 719–735 (2011)
7. Adamatzky, A.: On diversity of configurations generated by excitable cellular automata with dynamical excitation intervals. International Journal of Modern Physics C **23**(12), 1250085 (2012)
8. Adamatzky, A.: Reaction-diffusion automata. Springer (2012)
9. Adamatzky, A., De Lacy Costello, B., Asai, T.: Reaction-diffusion computers. Elsevier (2005)
10. Adamatzky, A., Mayne, R.: Actin automata: Phenomenology and localizations. International Journal of Bifurcation and Chaos **25**(02), 1550030 (2015)
11. Aizenman, M., Lebowitz, J.L.: Metastability effects in bootstrap percolation. Journal of Physics A: Mathematical and General **21**(19), 3801 (1988)
12. Alonso-Sanz, R.: Reversible cellular automata with memory: two-dimensional patterns from a single site seed. Physica D: Nonlinear Phenomena **175**(1), 1–30 (2003)
13. Alonso-Sanz, R.: Spatial order prevails over memory in boosting cooperation in the iterated prisoner's dilemma. Chaos: An Interdisciplinary Journal of Nonlinear Science **19**(2), 023102 (2009)
14. Alonso-Sanz, R.: Discrete systems with memory, vol. 75. World Scientific (2011)
15. Alonso-Sanz, R.: A glimpse of complex maps with memory. Complex Systems **21**(4) (2013)
16. Alonso-Sanz, R.: Scouting the Mandelbrot set with memory. Complexity **21**(3) (2016)
17. Alonso-Sanz, R.: On complex maps with delay memory. Fractals **23**(03), 1550027 (2015)
18. Alonso-Sanz, R., Martin, M.: Cellular automata with accumulative memory: legal rules starting from a single site seed. International Journal of Modern Physics C **14**(05), 695–719 (2003)
19. Alonso-Sanz, R., Martin, M.: Elementary cellular automata with elementary memory rules in cells: The case of linear rules. Journal of Cellular Automata **1**(1), 71–87 (2006)
20. Applegate, D., Pol, O.E., Sloane, N.J.A.: The toothpick sequence and other sequences from cellular automata. Congressus Numerantium **206**, 157–191 (2010)

21. Bandman, O.: Parallel simulation of asynchronous cellular automata evolution. In: Bandini, S., Yacoubi, S., Chopard, B. (eds.), Cellular Automata for Research and Industry (ACRI -2006). Lecture Notes in Computer Science, vol. 4173, pp. 41–47 (2006)

22. Bandman, O.: Cellular automata composition techniques for spatial dynamics simulation. In: Hoekstra, A.G., Kroc, J., Sloot, P.M.A., (eds.) Simulating Complex Systems by Cellular Automata, pp. 81–116 (2010)

23. Bandman, O.: Parallel composition of asynchronous cellualar automata simulating reaction diffusion processes. In: Cellular Automata in Research and Industry (ACRI–2010). Lecture Notes in Computer Science, vol. 6350, pp. 385–398 (2010)

24. Bandman, O.: Using multi core computers for implementing cellular automata systems. In: Malyshkin, V. (ed.) Parallel Computing Technologies(PaCt-2011). Lecture Notes in Computer Science, vol. 6873, pp. 140–151 (2011)

25. Bandman, O.: Implementation of large-scale cellular automata models on multi-core computers and clusters. In: 2013 International Conference on High Performance Computing and Simulation (HPCS), pp. 304–310. EEE Conference Publications (2013)

26. Berlekamp, E.R., Conway, J.H., Guy, R.K.: Winning Ways for your Mathematcial plays Volume 2: games in particular. Academic Press (1982)

27. Bussemaker, H.J., Deutsch, A., Geigant, E.: Mean-field analysis of a dynamical phase transition in a cellular automaton model for collective motion. Phys. Rev. Lett. **78**(26), 5018–5021 (1997)

28. Chaté, H., Manneville, P.: Collective behaviors in spatially extended systems with local interactions and synchronous updating. Progress of Theoretical Physics **87**(1), 1–60 (1992). Oxford University Press

29. Codd, E.F.: Cellular Automata. Academic Press, New York (1968)

30. Conti, C.: The enlightened Game of Life. In: Adamatzky, A., (ed.) Game of Life Cellular Automata. Springer (2002)

31. Cook, M.: A Concrete View of Rule 110 Computation. EPTCS **1**, pp. 31–55 (2009)

32. Cook, M.: Universality in elementary cellular automata. Complex Systems **15**(1), 1–40 (2004). Complex Systems Publications, Inc.

33. Cox, J.T., Griffeath, D.: Diffusive clustering in the two dimensional voter model. The Annals of Probability, 347–370 (1986)

34. Cox, J.T., Griffeath, D.: Recent results for the stepping stone model. In: Percolation Theory and Ergodic Theory of Infinite Particle Systems, pp. 73–83. Springer (1987)

35. Deutsch, A., Dormann, S.: Cellular Automaton Modeling of Biological Pattern Formation: Characterization, Applications, and Analysis. Birkhäuser, Boston (2005)

36. Devore, J., Hightower, R.: The Devore variation of the Codd self-replicating computer. In: Third Workshop on Artificial Life, Santa Fe, New Mexico, 1992. Original work carried out in the 1970s though apparently never published (reported by John R. Koza) [93]

37. Dewdney, A.K.: A cellular universe of debris, droplets, defects and demons. Scientific American **261**(2), 102–105 (1989)

38. d'Humières, D., Hasslacher, B., Lallemand, P., Pomeau, Y., Frisch, U., Rivet, J.-P.: Lattice gas hydrodynamics in two and three dimensions. Complex Systems **1**, 585–597 (1987)

39. Dong, C.Y.S., Long, J.T., Reiter, C.A., Staten, C., Umbrasas, R.: A cellular model for spatial population dynamics. Computers & Graphics **34**(2), 176–181 (2010)

40. Elokhin, V.I., Sharifulina, A.E.: Simulation ofheterogeneous catalytic reaction by asynchronous cellular automataon multicomputer. In: Malyshkin, V. (ed.) Parallel Computing Technologies(PaCt-2011). Lecture Notes in Computer Science, vol. 6873, pp. 204–209. Springer, Heidelberg (2011). (Kireeva)

41. Eppstein, D.: Holstein (b35678/s4678) (before 2007)

42. Evans, K.M.: Larger than Life: it's so nonlinear. PhD dissertation, University of Wisconsin-Madison (1996)

43. Evans, K.M.: Larger than Life: digital creatures in a family of two-dimensional cellular automata. In: Cori, R., Mazoyer, J., Morvan, M., Mosseri, R. (eds.) Discrete Mathematics and Theoretical Computer Science, AA, pp. 177–192 (2001)

44. Evans, K.M.: Larger than life: threshold-range scaling of Life's coherent structures. Physica D: Nonlinear Phenomena **183**(1), 45–67 (2003)
45. Evans, K.M.: Threshold-range scaling of Life's coherent structures. Physica D: Nonlinear Phenomena **183**(1–2), 45–67 (2003)
46. Evans, K.M.: Larger than Life's extremes: rigorous results for simplified rules and speculation on the phase boundaries. In: Adamatzky, A. (ed.) Game of Life Cellular Automata, pp. 179–221. Springer (2010)
47. Evans, K.M.: Larger than Life's invariant measures. Journal of Cellular Automata **6**(1), 53–75 (2011)
48. Fange, D., Elf, J.: Noise-induced min phenotypes in E. coli. PLoS Computational Biology **2**(6), 0637–0648 (2006)
49. Fatès, N.: Remarks on the cellular automaton global synchronisation problem. In: Kari, J. (ed.) AUTOMATA 2015 Proceedings of the 21st IFIP WG 1.5 International Workshopon Cellular Automata and Discrete Complex Systems. Lecture Notes in Computer Science, vol. 9099, pp. 113–126. Springer (2015)
50. Fatès, N., Morvan, M.: An experimental study of robustness to asynchronism for elementary cellular automata. Complex Systems **16**, 1–27 (2005)
51. Fatès, N., Morvan, M., Schabanel, N., Thierry, E.: Fully asynchronous behavior of double-quiescent elementary cellular automata. Theoretical Computer Science **362**, 1–16 (2006)
52. Fisch, R., Griffeath, D.: WinCA: A cellular automaton modeling environment (1996)
53. Fisch, R.: Cyclic cellular automata and related processes. Physica D: Nonlinear Phenomena **45**(1), 19–25 (1990)
54. Fisch, R., Gravner, J., Griffeath, D.: Cyclic cellular automata in two dimensions. In: Alexander, K.S., Wadkins, J.C. (eds.) Spatial Stochastic Processes. A festschrift in honor of the seventieth birthday of T.E. Harris. Birkhäuser, Boston (1991)
55. Fisch, R., Gravner, J., Griffeath, D.: Threshold-range scaling of excitable cellular automata. Statistics and Computing **1**(1), 23–39 (1991)
56. Fisch, R., Gravner, J., Griffeath, D.: Metastability in the Greenberg-Hastings model. The Annals of Applied Probability, 935–967 (1993)
57. Frisch, U., Hasslacher, B., Pomeau, Y.: Lattice-gas automata for the Navier-Stokes equation. Phys. Rev. Lett. **56**(14), 1505–1508 (1986)
58. Gale, D., Propp, J., Sutherland, S., Troubetzkoy, S.: Further travels with my ant. Mathematical Entertainments column, Mathematical Intelligencer **17**, 48–56 (1995)
59. Gardner, M.: Mathematical games: The fantastic combinations of John Conway's new solitaire game "life". Scientific American **223**(4), 120–123 (1970)
60. Georgoudas, G., Sirakoulis, G.C., Scordilis, E.M., Andreadis, I.: Parametric optimisation in a 2-d cellular automata model of fundamental seismic attributes with the use of genetic algorithms. Advances in Engineering Software **42**(9), 623–633 (2011)
61. Gerhardt, M., Schuster, H., Tyson, J.J.: A cellular automaton model of excitable media: Ii. curvature, dispersion, rotating waves and meandering waves. Physica D: Nonlinear Phenomena **46**(3), 392–415 (1990)
62. Gorodetskii, V.V., Elokhin, V.I., Bakker, J.W., Nieuwenhuys, B.E.: Field electron and field ion microscopy studies of chemical wave propagation in oscillatory reactions on platinum group metals. Catalysis Today **105**, 183–205 (2005)
63. Goucher, A.P.: Gliders in cellular automata on Penrose tilings. Journal of Cellular Automata **7**(5–6), 385–392 (2012)
64. Gravner, J., Griffeath, D.: Cellular automaton growth on Z^2: Theorems, examples, and problems. Advances in Applied Mathematics **21**, 241–304 (1998)
65. Gravner, J.: Cellular automata models of ring dynamics. International Journal of Modern Physics C **7**(06), 863–871 (1996)
66. Gray, G.: http://c2.com/cgi/wiki?DrCodd (2003) (accessed June 13, 2015)
67. Griffeath, D., Moore, C.: Life without Death is P-complete. Complex Systems **10**, 437–447 (1996)
68. Griffeath, D.: Self-organization of random cellular automata: four snapshots. In: Probability and Phase Transition, pp. 49–67. Springer (1994)

69. Griffeath, D.: Primordial Soup Kitchen (2006)
70. Griffeath, D., Moore, C.: Life without death is P-complete. Complex Systems **10**, 437–448 (1996)
71. Hartman, C., Heule, M., Kwekkeboom, K., Noels, A.: Symmetry in Gardens of Eden. Electr. J. Comb. **20**(3), P16 (2013)
72. Hartman, H., Tamayo, P.: Reversible cellular automata and chemical turbulence. Physica D: Nonlinear Phenomena **45**(1), 293–306 (1990)
73. Hatzikirou, H., Brusch, L., Schaller, C., Simon, M., Deutsch, A.: Prediction of traveling front behavior in a lattice-gas cellular automaton model for tumor invasion. Computers and Mathematics with Applications **59**(7), 2326–2339 (2010)
74. Hatzikirou, H., Voss-Böhme, A., Cavalcanti-Adam, E.A., Herrero, M.A., Deutsch, A., Böttger, K.: Emerging Allee effect in tumor growth. PLOS Comp. Bio. **11**(9), e1004366 (2015)
75. Hutton, T.: Video of the Munafo U-Skate glider on a Penrose P3 tiling (2014). http://www.youtube.com/watch?v=6JwsNdAlIog (accessed June 12, 2015)
76. Hutton, T., Munafo, R., Trevorrow, A., Rokicki, T., Wills, D.: Ready, a cross-platform implementation of various reaction-diffusion systems (2015). http://github.com/GollyGang/ready
77. Hutton, T.J.: Codd's self-replicating computer. Artificial Life **16**(2), 99–117 (2010)
78. Imai, K., Hatsuda, T., Poupet, V., Sato, K.: A universal semi-totalistic cellular automaton on kite and dart Penrose tilings (2012). arXiv preprint arXiv:1208.2771
79. Imai, K., Hori, T., Morita, K.: Self-reproduction in three-dimensional reversible cellular space. Artificial Life **8**(2), 155–174 (2002)
80. Kireeva, A.: A two-layer cellular automata model of carbon monoxide oxidation reaction. Bulletin of the Novosibirsk Computing Center, series Computer Science **36**, 33–46 (2014)
81. Korec, I.: Real-time generation of primes by a one-dimensional cellular automaton with 11 states. In: Mathematical Foundations of Computer Science 1997, pp. 358–367. Springer (1997)
82. Koza, J.R.: Artificial life: spontaneous emergence of self-replicating and evolutionary self-improving computer programs. In: Langton, C.G. (ed.) Artificial Life III, Proc. Volume XVII Santa Fe Institute Studies in the Sciences of Complexity, pp. 260. Addison-Wesley Publishing Company, New York (1994)
83. Krawczyk, R.J.: Experiments in architectural form generation using cellular automata. In: Koszewski, K., Wrona, S. (ed.) eCAADe20. Education in Computer Aided Architectural Design in Europe, Warsaw, Poland, pp. 552–555 (2002)
84. Krawczyk, R.J.: Experiments in architectural form generation using cellular automata. In: ecaade 2002 Conference, Warsaw, Poland, September 2002
85. Krawczyk, R.J.: Exploring the massing of growth in cellular automata. In: The 6th International Conference on Generative Art (2003)
86. Langton, C.G.: Studying artificial life with cellular automata. Physica D: Nonlinear Phenomena **22**, 120–149 (1986)
87. Langton, C.G.: Self-reproduction in cellular automata. Physica D: Nonlinear Phenomena **10**(1), 135–144 (1984)
88. Latkin, E.I., Elokhin, V.I., Gorodetskii, V.V.: Monte Carlo model of oscillatory co oxidation having regard to the change of catalytic properties due to the adsorbate-induced Pt(100) structural transformation. Journal of Molecular Catalysis A: Chemical **166**, 23–30 (2001)
89. Loose, M., Fischer-Friedrich, E., Herold, C., Kruse, K., Schwille, P.: Spatial regulators for bacterial cell division self-organize into surface waves in vitro. Science **320**, 789–792 (2008)
90. Loose, M., Fischer-Friedrich, E., Herold, C., Kruse, K., Schwille, P.: Min protein patterns emerge from rapid rebinding and membrane interaction of mine. Nature Structural & Molecular Biology **18**, 577–583 (2011)
91. Lutkenhaus, J.: Assebly dynamics of the bacterial mincde system and spatial regulation of the z ring. Annual Review of Biochemistry **76**, 539–562 (2007)
92. Makowiec, D.: Pacemaker rhythm through networks of pacemaker automata a review. Acta Phys. Pol. B Proc. Suppl. **7**(2)

93. Margenstern, M.: Cellular Automata in Hyperbolic Spaces: Theory. Volume 1, vol. 1. Archives contemporaines (2007)
94. Margenstern, M.: The domino problem of the hyperbolic plane is undecidable. Theoretical Computer Science **407**(1), 29–84 (2008)
95. Margenstern, M.: Bacteria, turing machines and hyperbolic cellular automata. A Computable Universe: Understanding and Exploring Nature as Computation (2012)
96. Margenstern, M.: About strongly universal cellular automata (2013). arXiv preprint arXiv:1304.6316
97. Margenstern, M.: A weakly universal cellular automaton in the pentagrid with five states. In: Computing with New Resources, vol. 8808, pp. 99–113. Springer (2014)
98. Margenstern, M.: Le rêve d'Euclide. In: Promenades En Géométrie Hyperbolique. Le Pommier (2015) (Euclid's dream, Walking in hyperbolic geometry, in French)
99. Markus, M., Hess, B.: Isotropic cellular automaton for modelling excitable media. Nature **347**(6288), 56–58 (1990)
100. Martínez, G.J.: Complex cellular automata repository (2010). http://uncomp.uwe.ac.uk/genaro/Complex_CA_repository.html (accessed August 06, 2015)
101. Martínez, G.J.: Rule 110 cellular automata repository (2010). http://uncomp.uwe.ac.uk/genaro/Rule110.html (accessed August 12, 2015)
102. Martínez, G.J., Adamatzky, A., Alonso-Sanz, R.: Designing complex dynamics in cellular automata with memory. International Journal of Bifurcation and Chaos **23**(10), 1330035 (2013). World Scientific
103. Martínez, G.J., Adamatzky, A., Chen, F., Chua, L.: On soliton collisions between localizations in complex elementary cellular automata: rules 54 and 110 and beyond. Complex Systems **21**(2), 117–142 (2012). Complex Systems Publications, Inc.
104. Martínez, G.J., Adamatzky, A., McIntosh, H.V.: A Computation in a Cellular Automaton Collider. In: Adamatzky, A. (ed.) Advances in Unconventional Computing. Springer (2016)
105. Martínez, G.J., Adamatzky, A., McIntosh, H.V.: Computing on rings. In: Zenil, H., Penrose, R. (eds) A Computable Universe: Understanding and Exploring Nature as Computation. World Scientific Press, pp. 257–276 (2012)
106. Martínez, G.J., McIntosh, H.V., Mora, J.C.S.T.: Gliders in rule 110. International Journal of Unconventional Computing **2**(1), 1 (2006). Old City Publishing Inc
107. Martínez, G.J., McIntosh, H.V., Mora, J.C., Vergara, S.V.C.: Determining a regular language by glider-based structures called phases fi_1 in Rule 110. Journal of Cellular Automata **3**(3), 231–270 (2008). Old City Publishing
108. Martínez, G.J., McIntosh, H.V., Mora, J.C.S.T., Vergara, S.V.C.: Reproducing the cyclic tag system developed by matthew cook with rule 110 using the phases f_1_1. Journal of Cellular Automata **6**(2–3), 121–161 (2011). Old City Publishing
109. Martínez, G.J., McIntosh, H.V., Mora, J.C.S.T., Vergara, S.V.C.: Rule 110 objects and other constructions based-collisions. Journal of Cellular Automata **2**(3), 219–242 (2007). Old City Publishing
110. Martínez, G.J., Adamatzky, A., Seck-Tuoh-Mora, J., Alonso-Sanz, R.: How to make dull cellular automata complex by adding memory: Rule 126 case study. Complexity **15**(16), 34–49 (2010). Wiley Online Library
111. Martínez, G.J., Adamatzky, A., Stephens, C.R., Hoeflich, A.F.: Cellular automaton supercolliders. International Journal of Modern Physics C **22**(04), 419–439 (2011)
112. Margenstern, M.: A weakly universal cellular automaton in the hyperbolic 3D space with three states (2010). arXiv preprint arXiv:1002.4290
113. McIntosh, H.V.: Linear cellular automata via de Bruijn diagrams (1991). http://delta.cs.cinvestav.mx/~mcintosh/cellularautomata/Papers_files/debruijn.pdf (accessed August 13, 2015)
114. McIntosh, H.V.: IX verano de investigación 1999 (1999). http://delta.cs.cinvestav.mx/~mcintosh/cellularautomata/Papers_files/summer99.pdf (accessed August 10, 2015)

115. McIntosh, H.V.: Rule110 as it rules relates to the presence of gliders (2002). `http://delta.cs.cinvestav.mx/~mcintosh/cellularautomata/Papers_files/rule110.pdf` (accessed August 05, 2015)
116. McIntosh, H.V.: One-dimensional cellular automata. Luniver Press, UK (2009)
117. Medvedev, Y.G.: Multi-particle cellular-automata models for diffusion simulation. In: Malyshkin, V., Hsu, C.-H. (eds.) Parallel Computing Technologies (PaCT-2009). Lecture Notes in Computer Science, vol. 6083, pp. 204–211 (2009)
118. Medvedev, Y.: Multiparticle CA model of fluid flow. Vestnik Tomskogo Gos. Universiteta, Upravlenie, Vychislitelnaya Tekhnika i Informatika 1(6) (2009)
119. Medvedev, Y.: Simulating a piston motion by a gas-lattice model. Prikladnaya Diskretnaya Matematika 4(10) (2010)
120. Mente, C., Prade, I., Brusch, L., Breier, G., Deutsch, A.: A lattice-gas cellular automaton model for in vitro sprouting angiogenesis. Acta Physica Polonica B 5(1), 99–115 (2012)
121. Minsky, M.L.: Computation: finite and infinite machines. Prentice-Hall Inc. (1967)
122. Mizas, C., Sirakoulis, G.C., Mardiris, V., Karafyllidis, I., Glykos, N., Sandaltzopoulos, R.: Reconstruction of dna sequences using genetic algorithms and cellular automata: Towards mutation prediction? Biosystems 92(1), 61–68 (2008)
123. Moore, E.F.: The firing squad synchronization problem, Selected Papers. In: Moore, E.F. (ed.) Sequential Machines. Addison-Wesley, Reading, MA (1964)
124. Moreira, J., Deutsch, A.: Cellular automaton models of tumor development: a critical review. Advances in Complex Systems 5(02n03), 247–267 (2002)
125. Munafo, R.: U-Skate World, an Instance of the Gray-Scott System (2015). `http://mrob.com/pub/comp/xmorphia/uskate-world.html` (accessed June 12, 2015)
126. Oliveira, G.M.B., Martins, L.G.A., de Carvalho, L.B., Fynn, E.: Some investigations about synchronization and density classification tasks in one-dimensional and two-dimensional cellular automata rule spaces. Electronic Notes in Theoretical Computer Science 252, 121–142 (2009)
127. Owens, N., Stepney, S.: The Game of Life rules on Penrose tilings: still life and oscillators. In: Adamatzky, A. (ed.) Game of Life Cellular Automata, pp. 331–378. Springer (2010)
128. Owens, N., Stepney, S.: Investigations of Game of Life cellular automata rules on Penrose tilings: Lifetime, ash, and oscillator statistics. Journal of Cellular Automata 5(3), 207–225 (2010)
129. Pennybacker, M., Newell, A.C.: Phyllotaxis, pushed pattern-forming fronts, and optimal packing. Phys. Rev. Lett. 110 (2013)
130. Pikovsky, A., Rosenblum, M., Kürths, J.: Synchronization. A Universal Concept in Nonlinear Sciences. Cambridge University Press (2001)
131. Reiter, C.A.: A local cellular model for snow crystal growth. Chaos, Solitons & Fractals 23(4), 1111–1119 (2005)
132. Reiter, C.A.: Cyclic cellular automata in 3D. Chaos, Solitons & Fractals 44(9), 764–768 (2011)
133. Rendell, P.: Turing universality of the Game of Life. In: Adamatzky, A. (ed.) Collision-Based Computing. Springer (2002)
134. Rendell, P.W.: Turing Machine Universality of the Game of Life. Springer (2015)
135. Rokicki, T., Trevorrow, A., Hutton, T., Greene, D., Summers, J., Verver, M., Munafo, R.: Golly, an open source, cross-platform application for exploring the Game of Life and other cellular automata (2015). `http://golly.sourceforge.net/`
136. Salzberg, C., Antony, A., Sayama, H.: Evolutionary dynamics of cellular automata-based self-replicators in hostile environments. BioSystems 78(1), 119–134 (2004)
137. Sayama, H.: Constructing evolutionary systems on a simple deterministic cellular automata space. PhD Dissertation, University of Tokyo, Department of Information Science (1998)
138. Sayama, H.: A new structurally dissolvable self-reproducing loop evolving in a simple cellular automata space. Artificial Life 5(4), 343–365 (1999)
139. Sayama, H.: Self-replicating worms that increase structural complexity through gene transmission. In: Artificial Life VII: Proceedings of the Seventh International Conference on Artificial Life, vol. 7, p. 21. MIT Press (2000)

140. Sayama, H.: An artificial life view of the Collatz problem. Artificial Life **17**(2), 137–140 (2011)
141. Seck-Tuoh-Mora, J.C., Chapa-Vergara, S.V., Martínez, G.J., McIntosh, H.: Procedures for calculating reversible one-dimensional cellular automata. Physica D **202**, 134–141 (2005)
142. Seck-Tuoh-Mora, J.C., Gonzalez-Hernandez, M., McIntosh, H., Chapa-Vergara, S.V.: Construction of reversible cellular automata by amalgamations and permutations of states. Journal of Cellular Automata **4**(3), 311–322 (2009)
143. Seck-Tuoh-Mora, J.C., Gonzalez-Hernandez, M., Perez-Lechuga, G.: An algorithm for analyzing the transitive behavior of reversible one-dimensional cellular automata with both Welch indices different. International Journal of Unconventional Computing **1**, 161–177 (2004)
144. Seck-Tuoh-Mora, J.C., Martínez, G.J., Alonso-Sanz, R., Hernández-Romero, N.: Invertible behavior in elementary cellular automata with memory. Information Sciences **199**, 125–132 (2012)
145. Seck-Tuoh-Mora, J.C., Martínez, G.J., Hernández-Romero, N., Medina-Marin, J.: Elementary cellular automaton rule 110 explained as a block substitution system. Computing **88**, 193–205 (2010)
146. Seck-Tuoh-Mora, J.C., Medina-Marin, J., Martínez, G.J., Hernández-Romero, N.: Emergence of density dynamics by surface interpolation in elementary cellular automata. Commun. Nonlinear Sci. Numer. Simulat. **19**, 941–966 (2014)
147. Sipper, M.: Evolution of Parallel Cellular Machines: The Cellular Programming Approach. Springer-Verlag, Heidelberg (1997)
148. Stewart, I.: A subway named Turing. Scientific American **271**, 104–107 (1994)
149. Turing, A.M.: The chemical basis of morphogenesis. Philosophical Transactions of the Royal Society of London **237**, 37–72 (1952)
150. Umeo, H.: Firing squad synchronization problem in cellular automata. In: Encyclopedia of Complexity and Systems Science, pp. 3537–3574. Springer (2009)
151. Umeo, H., Hisaoka, M., Akiguchi, S.: A twelve-state optimum-time synchronization algorithm for two-dimensional rectangular arrays. In: Unconventional Computation. LNCS, vol. 3699, pp. 214–223. Springer, Heidelberg (2005)
152. Umeo, H., Kamikawa, N.: Real-time generation of primes by a 1-bit-communication cellular automaton. Fundamenta Informaticae **58**(3–4), 421–435 (2003)
153. Umeo, H., Kamikawa, N., Nishioka, K., Akiguchi, S.: Generalized firing squad synchronization protocols for one-dimensional cellular automata - a survey. Acta Physica Polonica, B, Proceedings Supplement **3**(2), 267–289 (2010)
154. Umeo, H., Kamikawa, N., Yunès, J.-B.: A family of smallest symmetrical four-state firing squad synchronization protocols for ring arrays. Parallel Processing Letters **19**(02), 299–313 (2009)
155. Umeo, H., Kubo, K.: A seven-state time-optimum square synchronizer. In: Proc. of the 9th International Conference on Cellular Automata for Research and Industry. LNCS, vol. 6350, pp. 219–230 (2010)
156. Umeo, H., Miyamoto, K., Abe, Y.: Real-time prime generators implemented on small-state cellular automata. In: Adamatzky, A. (ed.) Automata, Universality, Computatopn, pp. 341–352. Springer (2015)
157. Umeo, H., Yanagihara, T.: Smallest implementations of optimum-time firing squad synchronization algorithms for one-bit-communication cellular automata. In: Parallel Computing Technologies, pp. 210–223. Springer (2011)
158. Ventrella, J.: A spherical xor gate implemented in the Game of Life. In: Game of Life Cellular Automata, pp. 379–385. Springer (2002)
159. Ventrella, J.: Glider dynamics on the sphere: Exploring cellular automata on geodesic grids. J. Cellular Automata **6**(2–3), 245–256 (2011)
160. Vitvitsky, A.A.: Cellular automata with a dynamical structure for simulating the biological tissues growth. Siberian Journal of Numerical Mathematics **17**(4), 315–327

161. Vitvitsky, A.A.: Construction of inhomogeneous 3D mesh for simulation of bacterial cell growth and division by cellular automata. Priklalnaya Diskretnaya Matematika **15**(3), 110–119 (2015)
162. Vitvitsky, A.: CA model of autowaves formation in the bacterial MinCDE system. In: Malyshkin, V. (ed.) Parallel Computing Technologies (PaCt-2015). Lecture Notes in Computer Science, vol. 9251, pp. 243–247 (2015)
163. von Neumann, J., Burks, W.: Theory of self-reproducing automata. Illinois University Press (1966)
164. Wolfram, S.: Cellular automata and complexity: collected papers. Addison-Wesley Publishing Company (1994)
165. Wolfram, S.: New Kind of Science. Wolfram Media Inc. (2002)
166. Wuensche, A.: Discrete dynamics lab (DDLab) (1993–2015). `http://www.ddlab.org/`
167. Wuensche, A.: Classifying cellular automata automatically. Complexity **4**(3), 47–66 (1999)
168. Wuensche, A.: Basins of attraction in cellular automata; order-complexity-chaos in small universes. Complexity **5**(6), 19–25 (2000)
169. Wuensche, A.: Glider dynamics in 3-value hexagonal cellular automata: The beehive rule. International Journal of Unconventional Computing **1**(4), 375–398 (2005)
170. Wuensche, A.: Exploring Discrete Dynamics. Luniver Press, UK (2011)
171. Wuensche, A., Lesser, M.: The Global Dynamics of Cellular Automata. Addison-Wesley, Reading (1992)
172. Yaguma, S., Odagiri, K., Takatsuka, K.: Coupled-cellular-automata study on stochastic and pattern-formation dynamics under spatiotemporal fluctuation of temperature. Physica D: Nonlinear Phenomena **197**(1), 34–62 (2004)
173. Yang, X.-S.: Computational modelling of nonlinear calcium waves. Applied Mathematical Modelling **30**(2), 200–208 (2006)
174. Young, D.A.: A local activator-inhibitor model of vertebrate skin patterns. In: Wolfram, S. (ed.) Theory and Applications of Cellular Automata. Advanced Series on Complex Systems, pp. 320–326 (1986)
175. Young, D.A.: A local activator-inhibitor model of vertebrate skin patterns. Math **72**(1), 51–58 (1984)

Index